The science and technology of materials in automotive engines

Related titles:

New developments in advanced welding
(ISBN-13: 978-1-85573-970-3; ISBN-10: 1-85573-970-4)
Recent technological developments have significantly transformed the welding industry. Automation, computers, process control, sophisticated scientific instruments and advanced processing methods are now common in today's industry. Engineers and technologists must support complex systems and apply sophisticated welding technologies. This comprehensive book discusses the changes in advanced welding technologies and prepares the reader for the modern industry.

Processes and mechanisms of welding residual stress and distortion
(ISBN-13: 978-1-85573-771-6; ISBN-10: 1-85573-771-X)
Measurement techniques for characterisation of residual stress and distortion have improved significantly. The development and application of computational welding mechanics have been phenomenal. Through the collaboration of a number of experts, this book provides a comprehensive discussion of the subject. It develops sufficient theoretical treatments on heat transfer, solid mechanics and materials behaviour that are essential for understanding and determining welding residual stress and distortion. It outlines an approach to computational analysis that engineers with sufficient background can follow and apply. This book will be useful for advanced analysis of the subject and provides examples and practical solutions for welding engineers.

Fundamentals of metallurgy
(ISBN-13: 978-1-85573-927-7; ISBN-10: 1-85573-927-5)
Part I of this book reviews the effects of processing on the properties of metals. A range of chapters cover such phenomena as phase transformations, types of kinetic reaction, transport and interfacial phenomena. The authors discuss how these processes and the resulting properties of metals can be modelled and predicted. Part II discusses the implications of this research for improving steelmaking and steel properties.

Details of these and other Woodhead Publishing materials books and journals, as well as materials books from Maney Publishing, can be obtained by:

- visiting our website at www.woodheadpublishing.com
- contacting Customer Services (e-mail: sales@woodhead-publishing.com; fax: +44 (0) 1223 893694; tel.: +44 (0) 1223 891358 ext. 30; address: Woodhead Publishing Ltd, Abington Hall, Abington, Cambridge CB1 6AH, England)

If you would like to receive information on forthcoming titles, please send your address details to: Francis Dodds (address, tel. and fax as above; email: francisd@woodhead-publishing.com). Please confirm which subject areas you are interested in.

Maney currently publishes 16 peer-reviewed materials science and engineering journals. For further information visit www.maney.co.uk/journals.

The science and technology of materials in automotive engines

Hiroshi Yamagata

**Woodhead Publishing and Maney Publishing
on behalf of
The Institute of Materials, Minerals & Mining**

**CRC Press
Boca Raton Boston New York Washington, DC**

WOODHEAD PUBLISHING LIMITED
Cambridge England

Woodhead Publishing Limited and Maney Publishing Limited on behalf of
The Institute of Materials, Minerals & Mining

Published by Woodhead Publishing Limited, Abington Hall, Abington
Cambridge CB1 6AH, England
www.woodheadpublishing.com

Published in North America by CRC Press LLC, 6000 Broken Sound Parkway, NW,
Suite 300, Boca Raton, FL 33487, USA

British Library Cataloguing in Publication Data
A catalogue record for this book is available from the British Library.

Library of Congress Cataloging in Publication Data
A catalog record for this book is available from the Library of Congress.

Woodhead Publishing Limited ISBN-13: 978-1-85573-742-6 (book)
Woodhead Publishing Limited ISBN-10: 1-85573-742-6 (book)
Woodhead Publishing Limited ISBN-13: 978-1-84569-085-4 (e-book)
Woodhead Publishing Limited ISBN-10: 1-84569-085-0 (e-book)
CRC Press ISBN-10: 0-8493-2585-4
CRC Press order number: WP2585

The publishers' policy is to use permanent paper from mills that operate a
sustainable forestry policy, and which has been manufactured from pulp
which is processed using acid-free and elementary chlorine-free practices.
Furthermore, the publishers ensure that the text paper and cover board used
have met acceptable environmental accreditation standards.

Project managed by Macfarlane Production Services, Markyate, Hertfordshire
(macfarl@aol.com)
Typeset by Replika Press Pvt Ltd, India
Printed by T J International Limited, Padstow, Cornwall, England

Contents

This book reviews the materials used in automotive engines. It discusses how the performance characteristics of engines are directly associated with the materials used and their methods of production.

This book has been written for those who are interested in automotive technologies, engineers and others who are engaged in the business of car parts and materials, and students who are learning mechanical engineering or materials engineering. The topics are centered on recent technologies as well as standards. Accordingly, this book will be a good introduction for those who intend to work in this field.

The 20th century became a society based on petroleum energy. The improved internal combustion engine can generate a high power output despite its small size, compared to the early engines at the end of the 19th century. Today's engines are used as power sources for various purposes including cars and motorcycles.

Modern cars use high technology in the materials field. The function of a component determines the materials to be used and their characteristics. A dialogue then takes place between the component designer and the materials manufacturer. The designer, for example, selects a shape designed to utilise fully the properties of the materials. The manufacturer chooses a production process to give a material its required properties. An experienced materials engineer can judge the technological sophistication of a manufacture by analysing the microstructure and chemical composition of the materials the manufacturer produces. The author has found that it is difficult for a beginner to enter this field. If there is a guidebook that links the functions of the engine to the material properties, it will assist the beginner to enter this field. This is the motive behind this book.

This book is arranged as follows. A brief explanation of engines as well as components is given in Chapter 1. Each following chapter gives the function and materials technology of an individual part of the engine. Appendix A is intended for the reader who has less knowledge of the basics of materials technology. The reader already having basic knowledge of processes such as quench hardening can understand the content easily. The reader who has not can also benefit because technical terms are explained when they appear for the first time.

The author acknowledges the book's dependence, directly and indirectly, on communications with the parts and materials manufacturers. I feel deep admiration for their efforts. Dr Graham Wylde of The Welding Institute is acknowledged for his proof reading and encouragement. Thanks are also due to my colleagues at Yamaha Motor Co., Ltd.

Hiroshi Yamagata

Note

Throughout the text '%' means 'wt%' unless otherwise indicated. The hardness value is shown as HB (Brinell hardness), HRB (Rockwell B scale), HRC (Rockwell C scale), HRF (Rockwell F scale) or HV (Vickers hardness). Please refer to a conversion manual if necessary.

1

Engines

1.1   The reciprocating engine

The engine is the heart of a car although it is normally hidden under the
bonnet. The engine is exposed in a motorcycle but the detailed mechanisms
are not visible. This chapter looks at these mechanisms.

Figure 1.1 shows a four-stroke cycle petrol engine with the various parts
indicated. In a reciprocating engine a mixture of petrol and air burns explosively

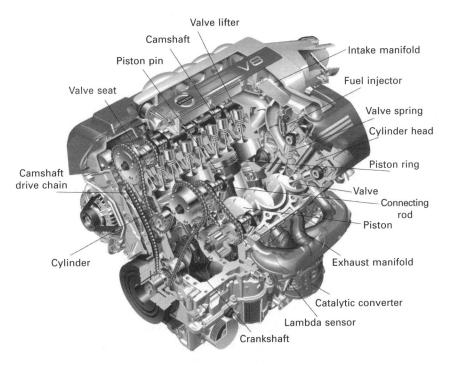

1.1 Cutaway of four-stroke cycle petrol engine (courtesy of Volvo Car
Corporation).

1

in a narrow container when ignited. The piston then receives the combustion pressure, and the connecting rod and crankshaft mechanism converts this pressure into rotation. This is the basic mechanism of a reciprocating engine. The reciprocating mechanism was originally inherited from steam engines and has been used for more than 200 years. One of the earliest mechanisms using a piston and cylinder can be seen in a 1509 drawing by Leonardo da Vinci, the famous painter and scientist of the Renaissance period. There are two main types of reciprocating engine, the four-stroke and the two-stroke engine. Figure 1.2 illustrates the sequence of operation. The four-stroke-cycle engine rapidly repeats strokes 1 to 4.[1, 2]

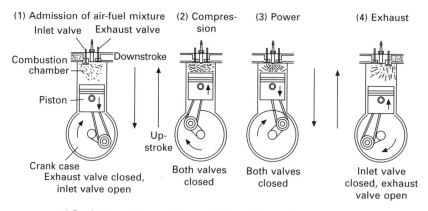

1.2 Basic operations of four-stroke cycle engine.

1.1.1   The four-stroke engine

The four-stroke engine is also referred to as the Otto cycle engine after its inventor N.A. Otto. Most cars use the four-stroke engine. An individual cycle comprises four strokes: 1, intake stroke; 2, compression stroke; 3, power stroke and 4, exhaust stroke. These four strokes repeat to generate the crankshaft revolution.

1. **Intake stroke**: the intake stroke draws air and fuel into the combustion chamber. The piston descends in the cylinder bore to evacuate the combustion chamber. When the inlet valve opens, atmospheric pressure forces the air-fuel charge into the evacuated chamber. As a result, the combustible mixture of fuel and air fills the chamber.
2. **Compression stroke**: at the end of the intake stroke, both inlet and exhaust valves are closed. The inertial action of the crankshaft in turn lifts the piston which compresses the mixture. The ratio of the combustion chamber volume before and after compression is called the compression ratio. Typically the value is approximately 9:1 in spark ignition engines and 15:1 in diesel engines.

3. **Power stroke**: when the piston ascends and reaches top dead center, an electric current ignites the spark plug and as the mixed gas burns, it expands and builds pressure in the combustion chamber. The resulting pressure pushes the piston down with several tons of force.
4. **Exhaust stroke**: during the exhaust stroke, the inlet valve remains closed whilst the exhaust valve opens. The moving piston pushes the burned fumes through the now open exhaust port and another intake stroke starts again.

During one cycle, the piston makes two round trips and the crankshaft revolves twice. The inlet and exhaust valves open and close only once. The ignition plug also sparks only once. A petrol engine, whether four- or two-stroke, is called a spark ignition (SI) engine because it fires with an ignition plug. The four-stroke-cycle engine contains the lubricating oil in the crankcase. The oil both lubricates the crankshaft bearings and cools the hot piston.

1.1.2   The two-stroke engine

The two-stroke engine is similar to that of the four-stroke-cycle engine in its reciprocating mechanism. It uses the piston-crankshaft mechanism, but requires only one revolution of the crankshaft for a complete power-producing cycle. The two-stroke engine does not use inlet and exhaust valves. The gas exchange is implemented by scavenging and exhaust porthole openings in the bore wall. The upward and downward motion of the piston simultaneously opens and closes these portholes. The air-fuel mixture then goes in or out of the combustion chamber through the portholes. Combustion takes place at every rotation of the crankshaft.

In the two-stroke engine, the space in the crankcase works as a precompression chamber for each successive fuel charge. The fuel and lubricating oil are premixed and introduced into the crankcase, so that the crankcase cannot be used for storing the lubricating oil. When combustion occurs in the cylinder, the combustion pressure compresses the new gas in the crankcase for the next combustion. The burnt gas then exhausts while drawing in new gas. The lubricating oil mixed into the air-fuel mixture also burns.

Since the two-stroke engine does not use a valve system, its mechanism is very simple. The power output is fairly high because it achieves one power stroke per two revolutions of the crankshaft. However, although the power output is high, it is used only for small motorcycle engines and some large diesel applications. Since the new gas pushes out the burnt gas, the intake and exhaust gases are not clearly separated. As a result, fuel consumption is relatively high and cleaning of the exhaust gas by a catalytic converter is difficult.

In the past, petrol engines almost universally used[3] a carburetor. However, the requirements for improved fuel economy have led to an increasing use of fuel injection. In a petrol engine the fuel is normally injected into the inlet manifold behind the inlet valve. The atomized fuel mixes with air. When the inlet valve is opened, the combustible mixture is drawn into the cylinder. However, a recent development has occurred in direct injection petrol engines whereby fuel is injected directly into the combustion chamber, as with direct injection diesel engines.

1.1.3   The diesel engine

The name diesel comes from the inventor of the diesel engine, R. Diesel. There are both four- and two-stroke-cycle diesel engines. Most automotive diesels are four-stroke engines. The intake stroke on the diesel engine draws only air into the cylinder. The air is then compressed during the compression stroke. At near maximum compression, finely atomized diesel fuel (a gas oil having a high flashpoint) is sprayed into the hot air, initiating auto ignition of the mixture. During the subsequent power stroke, the expanding hot mixture works on the piston, then burnt gases are purged during the exhaust stroke. Since diesel engines do not use a spark plug, they are also referred to as compression ignition (CI) engines. In the case of petrol engines, too high a temperature in the combustion chamber ignites the petrol spontaneously. When this occurs, the plug cannot control the moment of ignition. This unwanted phenomenon is often referred to as 'knocking'.

The diesel is an injection engine. A petrol engine normally needs a throttle valve to control airflow into the cylinder, but a diesel engine does not. Instead, the diesel uses a fuel injection pump and an injector nozzle sprays fuel right into the combustion chamber at high pressure. The amount of fuel injected into the cylinder controls the engine power and speed. There are two methods[3] by which fuel is injected into a combustion chamber, direct or indirect injection. With direct injection engines (DI) the fuel is injected directly into the cylinder and initial combustion takes place within the bowl that is machined into the piston head itself. With indirect injection engines (IDI) the fuel is injected and initial combustion takes place in a small pre-combustion chamber formed in the cylinder head. The burning gases then expand into the cylinder where combustion continues. Pistons for IDI engines usually have shallow depressions in their heads to assist the combustion process. Although an IDI engine has some advantages, it cannot match the efficiency of a DI engine, which is why most new automotive diesel engines entering production are DI designs.

Turbo-charged engines are mainly used because diesels can generate only a low power output without turbocharging. Turbocharging with an intercooler is used in large engines. Diesel engines produce lean combustion, having an air-fuel ratio of about 15:1 up to 100:1. The diesel's leaner fuel mixture generates higher fuel economy compared to that of a petrol engine. The peak cylinder pressure can be in excess of 15 MPa. The HC and CO contents in the exhaust gas are lower compared to those of petrol engines, but the particulate soot and NOx emissions cause environmental problems. In comparison with petrol engines, the components in a diesel engine are exposed to significantly more arduous operating conditions. Up until the 1980s, the noise, exhaust smoke and poor performance of diesel engines made them less attractive. However, recently improved diesel engines with high torque now offer a more attractive alternative to petrol engines.

A Stirling engine is another type of engine that uses a piston-cylinder construction. There are, however, other engines, such as the rotary and gas turbine engines, that do not use the piston-cylinder mechanism.

1.2   Advantages and disadvantages of reciprocating engines

An engine with a piston-cylinder mechanism has the following advantages:

1. It is possible to seal the gap between the piston and the cylinder, resulting in high compression ratio, high heat efficiency and low fuel consumption.
2. The piston ring faces the cylinder bore wall, separated by an oil film. The resulting hydrodynamic lubrication generates low friction and high durability.
3. The piston loses speed at the dead-center points where the travelling direction reverses, which gives enough time for combustion and intake as well as for exhaust.

However, the reciprocating engine also has disadvantages:

1. The unbalanced inertial force and resulting piston 'slap' can cause noise and vibration.
2. It is difficult to reuse the exhaust heat.

The rotary engine (Wankel engine) is one of the few alternatives that have been mass produced and installed in production vehicles. However, none of them has been as popular as the piston-cylinder mechanism to date.

1.3   Engine components and typical materials

1.3.1   Components

The reciprocating engine generates rotation from combustion pressure using the piston, connecting rod and crankshaft. Now let us look at the cutaway image of the four-stroke cycle petrol engine in Fig. 1.1 and its actual parts in Fig. 1.3. The basic functions of the various parts are as follows: the piston receives combustion pressure; the connecting rod transmits the combustion pressure to the crankshaft and the crankshaft transforms this reciprocating

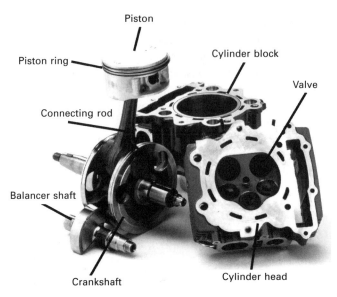

1.3 Parts for a single-cylinder four-stroke engine.

motion into smooth rotation. The combustion of the air-fuel mixture takes place in the chamber formed between the piston and the cylinder head. The piston then moves up and down in the cylinder via combustion pressure and the combustion gas is then sealed by the piston ring, which contains the pressure. The four-stroke engine requires a valve system that takes in and exhausts the combustion gas. For effective combustion, it is very important that the valve, ring and other components do not allow any leakage of pressure from the combustion chamber.

Figure 1.1 illustrates an engine containing four valves per cylinder. The camshaft pushes the valves into the combustion chamber. The repulsive force of the valve spring drives the backward motion. The valves and valve seats fitted in the cylinder head then seal the combustion gas. The following chapters will discuss the main components of the engine in more detail. As the mechanism for generating power, the cylinder, piston, and piston ring are discussed first. The mechanism that controls combustion includes the camshaft, valve, valve seat and valve spring. These components are discussed next. The mechanism that transforms reciprocating motion to rotation uses the crankshaft and connecting rod. Finally, the catalyst, turbocharger and exhaust manifold are presented as the mechanism for dealing with the exhaust gas.

1.3.2   Typical materials

Table 1.1 lists the typical metals used in engine parts. Metals such as iron (Fe), lead (Pb), and tin (Sn), are all mixed to bring out various properties.

Table 1.1 Typical metals for engine parts

Part name	Material
Cylinder block	Gray cast iron, compact graphite cast iron, cast Al alloy
Piston	Al-Si-Cu-Mg alloy
Piston ring	Gray cast iron, spheroidized graphite cast iron, alloy cast iron, spring steel and stainless steel
Camshaft	Chilled cast iron, Cr-Mo steel, iron base sintered metal
Valve	Heat-resistive steel, Ti alloy, SiC ceramics
Valve seat	Iron base sintered metal, cast iron
Valve spring	Spring steel, music wire
Piston pin	Nodular cast iron, Si-Cr steel, stainless steel
Connecting rod	Carbon steel, iron base sintered metal, micro-alloyed steel, spheroidized graphite cast iron
Crankshaft	Carbon steel, micro-alloyed steel, Cr-Mo steel and nodular cast iron
Turbo charger	Niresist cast iron, cast stainless steel, superalloy
Exhaust manifold	High-Si cast iron, niresist cast iron, cast stainless steel, stainless steel tube and sheet
Plain bearing	Al-Si-Sn and Cu-Pb alloys
Catalyst	Pt-Pd-Rh alloy

Statistics for the year 2000 state that the ratio of materials in cars is: steel plate 37%, steel bar 23%, cast iron 8%, aluminum alloy 8%, other non-ferrous alloys 2%, plastics 10%, rubber 7%, glass 2% and others 3%. The recent trend to pursue more lightweight materials has also reduced the ratio of steel. However, the main materials used for engine parts are iron base alloys such as structural steels, stainless steels, iron base sintered metals, and cast iron and aluminum alloy parts for the piston, cylinder head and cylinder block.

1.4   Recent trends in engine technology

In the earlier section, we looked at the structures, parts, and materials. Now, let us look at recent trends in engine technology. To illustrate performance development, Fig. 1.4[4] provides a comparison of the various engine designs used for car and commercial vehicle power units. Petrol engines have developed the following technologies

1. The multi-valve engine was previously limited to sports cars and motorcycles. To obtain higher output power, the number of valves used in car engines has increased.

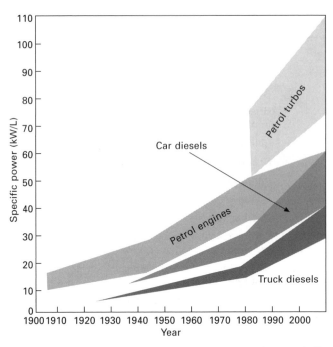

1.4 Development of power output of petrol engines and diesel
engines.

2. The multi-cylinder engine has become more widespread. It has a smoother
    rotation to decrease noise and vibration.
3. Three-way catalyst (Pt-Pd-Rh alloy) technology, using O_2 and knock
    sensors, has decreased the three components CO, HC, and NOx in the
    exhaust gas, to decrease environmental pollution.
4. The variable valve system has decreased fuel consumption.
5. Decreased inertial weight and electronic control have given improved
    engine performance.
6. Hybrid systems including an electric motor have reduced fuel consumption.

At the end of the 20th century, automotive diesel technology has made
significant progress. Diesel engines play a significant role[4] in reducing fuel
consumption in cars. In Europe, where this is already an important issue in
contrast to the United States or Japan, current estimates indicate that the
development target of a '3-litre car' can only be implemented with a diesel
engine. The diesel output power for a passenger car as well as for big
commercial vehicles is likely to increase. As shown in Fig. 1.4, the specific
power output range of diesels now equals that of naturally aspirated petrol
engines. The average output power is 50 kW/L. Most of the direct-injection
diesel engines for cars around the year 2003 have reached specific power

outputs of up to about 60 kW/L. Diesel cars are now available with power outputs to 180 kW and levels of low speed torque that were previously unimaginable. A truck using a large diesel engine has a maximum power around 400 kW.

These engines are characterized by four valves per cylinder, a combustion bowl located centrally in the piston and second-generation high-pressure injection systems (common rail, unit injector) with efficient control of the injection process by electronic means. The common rail technology provides better fuel efficiency, and better torque at low speeds. The increase in performance is combined with an increase in cylinder pressures up to peak values of 18 MPa.

The additional air supply by exhaust gas circulation also offers the possibility of very high specific power outputs, which will stretch the performance and fuel economy boundaries even further. The more stringent NOx limit requires additional after-treatment technology, and particulate filters will become the norm for diesel engine applications in cars.

The use of diesel engines remains on a continuing upward trend. The reason behind the explosion in the European market for diesel engines is generally to do with rapid advances in technology rather than simple fuel economy. Technology has now given diesel engines high performance and favorable torque characteristics.

1.5   References and notes

1. Duffy J.E., *Auto Engines*, New York, The Goodheart-Willcox Company, Inc. (1997).
2. *Automotive Handbook*, 5th edition, ed. by Bauer H. Warrendale SAE, Society of Automotive Engineers, (2000).
3. Federal-Mogul corporate catalogue, (2003).
4. Kolbenschmidt and Pierburg, Homepage, http://www.kolbenschmidt-pierburg.com (2003).

2

The cylinder

2.1   Structures and functions

The cylinder block is the basic framework of a car engine. It supports and holds all the other engine components. Figure 2.1 shows a typical cylinder block without an integrated crankcase. Figure 2.2 shows the block with the upper part of the crankcase included. Figure 2.3[1] schematically illustrates the relative positions of the cylinder, piston and piston ring. The cylinder is a large hole machined in the cylinder block, surrounded by the cylinder wall. The piston rapidly travels back and forth in the cylinder under combustion pressure. The cylinder wall guides the moving piston, receives the combustion pressure, and conveys combustion heat outside the engine. Figure 2.4 gives an analysis of the materials needed for a cylinder with high output power and

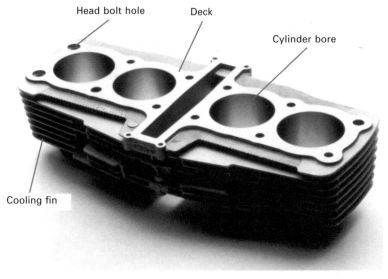

Head bolt hole   Deck

Cylinder bore

Cooling fin

2.1 Air-cooled block.

2.2 Cast iron cylinder block (closed deck type) including a crankcase portion.

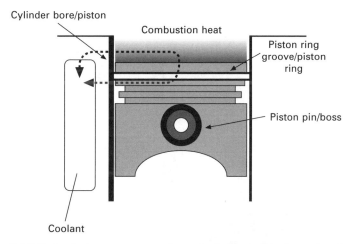

2.3 Tribological system around a cylinder bore (black portions). These are: the running surfaces between the piston pin and piston boss, between the cylinder bore and piston, and the piston ring groove and piston ring.

summarizes the reasons why a specific material or technology is chosen to fulfil a required function. A more detailed description is given in Appendix B.

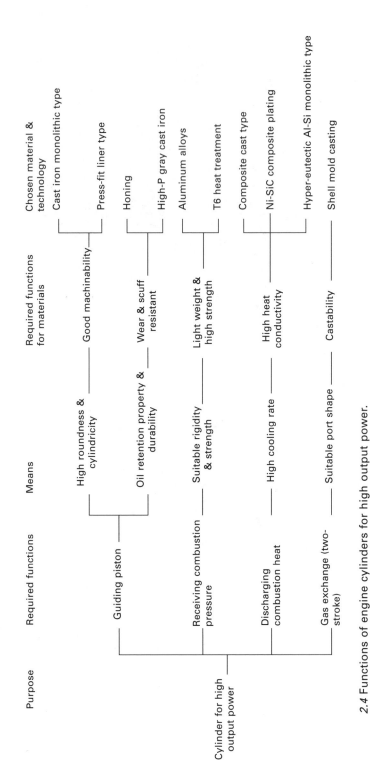

2.4 Functions of engine cylinders for high output power.

The black portions in Fig. 2.3 indicate the areas that are most exposed to friction. These parts need to be carefully designed not only from the viewpoint of lubrication but also tribology, as this has a significant influence on engine performance. Tribology can be defined as the science and technology of interacting surfaces in relative motion, and includes the study of friction, wear and lubrication. Combustion heat discharges at a very high rate and, if not diffused, the raised temperature can lead to tribological problems.

The cylinder must maintain an accurate roundness and straightness of the order of μm during operation. The cylinder bore wall typically experiences local wear at the top-dead-center point, where the oil film is most likely to fail, and scratching along the direction of travel of the piston. Figure 2.5 shows vertical scratching caused by scuffing. The grooves caused by scratching increase oil consumption and blow-by. In extreme cases, the piston seizes to the bore wall. The demand for higher output with improved exhaust gas emission has recently increased heat load to the cylinder even more. A much lighter weight design is also required.

2.5 Scuffing at bore surface; piston travelling vertically.

An engine generating high power output requires more cooling, since it generates more heat. Automotive engines have two types of cooling systems, air-cooled and water-cooled. Figure 2.1 shows the air-cooled type and Fig. 2.2 the water-cooled type. Whilst an air-cooled engine may use a much simpler structure because it does not use the water-cooled system, the heat management of the cylinder block is not as easy. As a result, most automotive engines nowadays use water-cooled systems. It would be no exaggeration to say that the required cooling level for an individual engine determines its cylinder structure.

Figure 2.6 shows cutaway views of four different types of cylinder block structure. The monolithic or quasi-monolithic block (on the right) is made of only one material. It is also called a linerless block because it does not contain liners (described later). The bore wall consists of either the same material as the block or a modified surface such as plating to improve wear resistance. It is normally difficult for one material to fulfill the various needs listed in Fig. 2.4. However a liner-less design in multi-bore engines can make the engine more compact by decreasing inter-bore spacing.

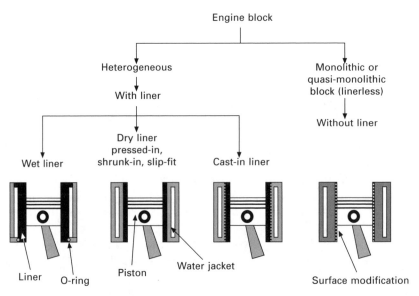

2.6 Bore designs in engine blocks.

The other designs in Fig. 2.6 (on the left) incorporate separate liners. A liner is also called a sleeve. A wet liner is directly exposed to coolant at the outer surface so that heat directly dissipates into the coolant. To withstand combustion pressure and heat without the added support of the cylinder block, it must be made thicker than a dry liner. A wet liner normally has a flange at the top. When the cylinder head is installed, the clamping action pushes the liner into position. The cylinder head gasket keeps the top of the liner from leaking. A rubber or copper O-ring is used at the bottom, and sometimes at the top, of a wet liner to prevent coolant from leaking into the crankcase. A dry liner presses or shrinks into a cylinder that has already been bored. Compared to the wet liner, this liner is relatively thin and is not exposed to the coolant. The cast-in liner design encloses the liner during the casting process of an entire cylinder block.

Table 2.1 lists various types of cylinder structures, their processing and characteristics. Cylinder blocks are normally made of cast iron or aluminum alloy. The aluminum block is much lighter. Various types of materials are combined to increase strength. In the following sections, we will look at the blocks of four-stroke engines. Those for two-stroke engines are discussed in the final section.

2.2   The cast iron monolithic block

The use of cast iron blocks in Table 2.1 has been widespread due to low cost as well as formability. Figure 2.2 shows a V6 block used for a car engine. The block is normally the integral type where the cylinders and upper crankcase are all one part. The cylinders are large holes that are machined into the block. The iron for the block is usually gray cast iron having a pearlite-microstructure, typically being JIS-FC200 (Table 2.2). The microstructure is shown in Fig. 2.7. Gray cast iron is so called because its fracture has a gray appearance. Ferrite in the microstructure of the bore wall should be avoided because too much soft ferrite tends to cause scratching, thus increasing blow-by.

Cast iron blocks are produced by sand casting. For cast iron, the die casting process using a steel die is fairly rare. The lifetime of the steel die is not adequate for repeated heat cycles caused by melting iron. As its name suggests, sand casting uses a mold that consists of sand. The preparation of sand and the bonding are a critical and very often rate-controlling step. Permanent patterns are used to make sand molds. Generally, an automated molding machine installs the patterns and prepares many molds in the same shape. Molten metal is poured immediately into the mold, giving this process very high productivity. After solidification, the mold is destroyed and the inner sand is shaken out of the block. The sand is then reusable. Two main methods are used for bonding sand. A green sand mold consists of mixtures of sand, clay and moisture. A dry sand mold consists of sand and synthetic binders cured thermally or chemically.

Figure 2.8 shows a schematic view of a sand mold used to shape a tube. This mold includes a sand core to make the tube hollow. The casting obtained from using this mold is shown in Fig. 2.9. Normally, molten iron in a ladle is gently poured into the cavity under the force of gravity using a filling system. The sand core forming an inside hollow shape is made from a dry sand component. The bore as well as the coolant passages in the cylinder block are shaped as cored holes.

Table 2.1 Cylinder structures

Type	Structure	Processing	Characteristics
Monolithic (linerless)	(1) Cast iron integrated type.	Monolithic block (typically, JIS-FC 200) with sand casting. The water passage is formed using expendable shell core. Laser or induction hardening is sometimes used on the bore surface to give durability.	Low cost but heavy.
Heterogeneous (dry liner)	(2) Cast iron block enclosing cast iron liner.	High-P cast iron liner is slip-fitted in JIS-FC200 block.	Hard liner gives durability.
Heterogeneous (cast-in liner)	(3) Aluminum block enclosing cast iron liner.	Liner is enclosed in block (typically, JIS-ADC12 die casting, JIS-AC4B shell molding) by casting-in with various casting methods.	Better cooling performance than type (1).
Heterogeneous (cast-in liner)	(4) Aluminum block enclosing PM-aluminum liner.	PM aluminum liner is enclosed in block (typically, JIS-ADC12 diecasting) by casting-in with high-pressure die casting.	Better cooling performance than type (3).
Heterogeneous (dry liner)	(5) Aluminum block enclosing cast iron or hyper-eutectic Al-Si liner with press-fitting.	Liner is inserted in block (typically, JIS-ADC12 die casting, JIS-AC4B shell molding) by press-fitting or shrunk-in.	Accurate roundness at elevated temperatures.
Quasi-monolithic (linerless)	(6) Aluminum block with plated bore surface.	Monolithic block having a coated bore by porous-Cr or Ni-SiC plating. The block material is typically JIS-AC4B shell molding or JIS-ADC12 high-pressure die casting.	High cooling performance. Bore pitch can be shortened in multi-bore engines.
Quasi-monolithic (linerless)	(7) Aluminum block with metal-sprayed bore surface.	Wire explosion or plasma spraying (steel base alloy) on the aluminum bore wall.	Cooling performance is the same as (6).
Monolithic (linerless)	(8) Hyper-eutectic Al-Si block without coating.	Low-pressure die casting using A390 alloy. The bore surface is either etched or mechanically polished to expose Si.	The wear-resistant coating is necessary on the piston side.
Quasi-monolithic (linerless)	(9) Fiber or particle rein-forced Al alloy composite.	Preform of fibers (Saphire+carbon) or Si particle is cast into aluminum by squeeze die casting.	The rigidity of the cylinder bore increases.

2.7 Microstructure of a gray cast iron (JIS-FC200) block.

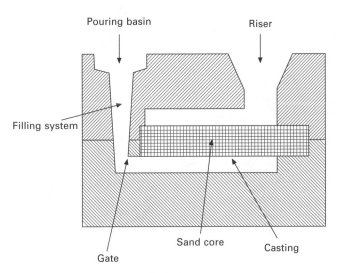

2.8 Sand mold with a sand core.

2.2.1   Honing, lubrication and oil consumption

Cast iron has been successfully used for monolithic blocks. This is because the casting process can mass-produce large complex shapes. The cylinder bore must have high dimensional accuracy. Honing is the finishing process

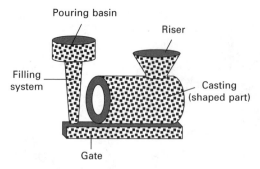

2.9 Cylindrical casting obtained using the mold shown in Fig. 2.8.

used to give accurate roundness and straightness. It is performed after the fine boring process. Figure 2.10 shows a micrograph of a honed bore surface. The honing whetstone carved the crosshatch pattern. During engine operation, the groove of the crosshatch holds lubricating oil. The resulting oil film generates hydrodynamic lubrication.

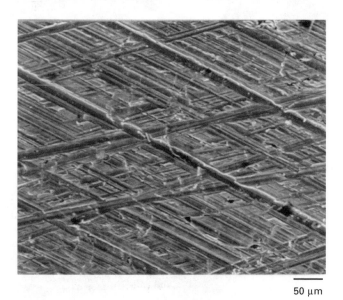

50 μm

2.10 Crosshatch pattern after honing.

Figure 2.11 is a picture of a honing machine. The tool shaft installs a honing head with a segmented whetstone at its end (Fig. 2.12). The whetstone grinds the bore by exerting an expanding pressure. The vertical motion of the head together with revolution generates the crosshatch pattern. The sharpness of the whetstone determines the profile of the crosshatch. A dull honing

2.11 Honing machine. The honing head installing whetstone is hanging from the top. The jig fixes the cylinder block. The segmented whetstone around the head expands to hone the bore. Experience is necessary in choosing the whetstone.

2.12 Honing tool head.

stone with an excess pressure gives an unwanted over-smeared surface. Ideally the finished surface exposes graphite without burr. The quality of the honing is measured by surface roughness value.

The graphite in the cast iron block works as a solid lubricant during machining as well as in engine operation. A solid lubricant gives a low frictional force without hydrodynamic lubrication. Graphite, MoS_2, WS_2, Sn, and Pb are all well known as solid lubricants. The low frictional force of graphite comes from the fact that the crystal structure has a very low frictional coefficient during slip at the basal plane. Figure 2.13 shows a schematic representation of the mechanism. The crystal slides easily along its basal plane even with a low shear force. The graphite decreases friction for tools during machining. The brittle nature of graphite makes chips discontinuous. The resultant high machinability gives high dimensional accuracy to cast iron parts. The graphite also works as a solid lubricant to prevent seizure of the piston or piston ring even under less oily conditions.

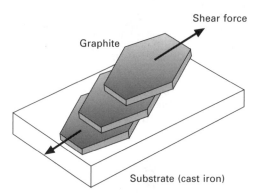

2.13 Mechanism of solid lubrication. The graphite in cast iron deforms like this under frictional force. Graphite has a layered crystal structure. A low friction coefficient appears along its basal plane under shear stress.

The micro-burr of the crosshatch disrupts the oil film to obstruct hydrodynamic lubrication. Additional Mn-phosphate conversion coating (refer to Appendix H) chemically removes the micro-burr to increase oil retention. This prevents seizure during the running-in stage. As well as dimensional accuracy, the surface profile also determines oil retention which, in turn, greatly influences wear resistance. An appropriate profile should be established. One pass finish with the whetstone usually shapes the surface profile to the normal type shown in Fig. 2.14(a)[2]. An additional finish, scraping off the peak, generates the trapezoid pattern shown in Fig. 2.14(b). This finishing is called plateau honing.

A customer does not want to change engine oil frequently. Less oil consumption is therefore required. Figure 2.15[2] compares the oil consumption

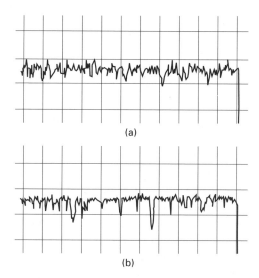

2.14 Two types of honed surfaces: (a) normal type; (b) plateau type. One graduation of the vertical axis measures 1 μm, that of the horizontal axis 0.1 mm (adapted from ref. 2).

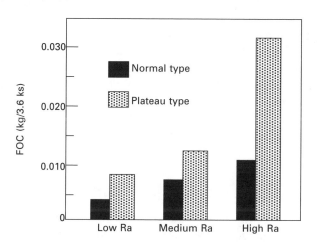

2.15 Comparison of oil consumption in honed shapes.

of a 1.9L car engine, measured by the final oil consumption value (FOC). For a normal type profile, the low Ra = 0.12 μm, middle Ra = 0.4 μm and high Ra = 0.62 μm. For the plateau type, the low Ra = 0.14 μm, middle Ra = 0.32 μm and high Ra = 0.88 μm. The oil consumption is least in the normal type of low Ra. However, it is worth mentioning that the scuffing resistance of the low Ra surface is poor. When the bore wall temperature is high, the plateau surface shows excellent resistance to scuffing although oil consumption is high. This feature comes from the fact that the plateau shape can maintain more lubricating oil without disrupting the oil film.

2.2.2   Improvement of wear resistance of cast iron blocks

Four-stroke engines splash oil on the bore wall for lubrication and cooling. To scrape off the excess oil, the contact pressure of the oil control ring is set high.[3] This decreases oil consumption. Hence, the bore wall should have higher wear resistance. To raise durability, a hard gray cast iron containing phosphorus (P) is often used (high-P cast iron in Table 2.2). Figure 2.16 shows the microstructure of high-P cast iron. The increased P crystallizes from the melt as hard steadite. It has a chemical composition of Fe_3P. The curious shape of steadite stems from its low freezing point. The iron crystal solidifies first. Then, the residual liquid solidifies to form steadite in the space between the iron crystals. This alloy composition has good wear resistance because of high hardness, but low machinability. Hence, instead of using this composition for an entire block, it is typical to enclose it as a wet or dry liner in a normal cast iron block (2 in Table 2.1) or aluminum block (3 and 5 in Table 2.1, described later).

The mileage required for commercial diesel engines is very high, being as much as 1,000,000 km. These engines have high combustion temperatures. Engines requiring very long durability use additional heat treatment on the bore surface. A nitrided liner[4] is often enclosed to increase hardness. A phosphate conversion coating on the liner also prevents corrosion. Instead of enclosing a hard liner, interrupted quench hardening by laser or induction heating can also be applied to the bore wall of the monolithic cast iron block.[4, 5]

2.3   The compact graphite iron monolithic block

There have been trials to improve the strength of gray cast iron without losing its superior properties. Petrol engines have cylinder pressures ranging from 7 to 12 MPa, while heavy-duty diesels operate in excess of 20 MPa. This high pressure generates much higher mechanical and thermal stresses on the cylinder block. The use of a cast iron block is widespread because of the high strength needed. However, a much stronger material is required to enable a lightweight design with decreased thickness. For these requirements, a block made of compact graphite iron has been proposed.

The graphite shape greatly influences the material characteristics of cast iron. Figure 2.17 gives a schematic representation of graphite morphology. In the conventional casting procedure, cast iron generates a flaky shape (a) in the iron matrix. However, when a special modification treatment is

Table 2.2 Chemical compositions (%). JIS-FC200 is a flake graphite cast iron having a strength of 200 MPa. JIS-AC4B and ADC12 are aluminum alloy for castings. A 390 is a hyper-eutectic Al-Si alloy

Material	Si	Fe	Cu	Mn	Mg	Zn	Ni	Ti	Cr	Al	C	P	V
JIS-FC200	2.0	Balance	–	0.8	–	–	–	–	–	–	3.2	–	–
High-V cast iron	2.0	Balance	–	0.8	–	–	–	–	–	–	3.2	–	0.3
High-P cast iron	2.0	Balance	–	0.8	–	–	–	–	–	–	3.2	0.3	0.3
JIS-AC4B	8.0	1.0	0.3	0.4	0.3	0.5	0.1	0.2	0.1	Balance	–	–	–
JIS-ADC12	11.0	1.3	2.0	0.5	0.3	1.0	0.5	–	–	Balance	–	–	–
A390	18.0	0.5	4.0	0.1	0.5	–	–	0.2	–	Balance	–	–	–

40 μm

2.16 High-P cast iron. The matrix has a pearlite microstructure, being a mixture of cementite and ferrite. Steadite is a eutectic crystal containing Fe_3P compound and ferrite. It appears typically in the cast iron of 0.3% P and over. See Appendix D.

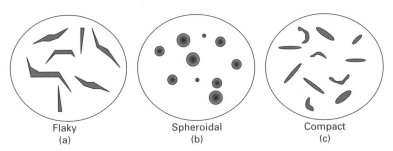

| Flaky | Spheroidal | Compact |
| (a) | (b) | (c) |

2.17 Graphite shape in cast iron: (a) flaky graphite; (b) spheroidal (nodular) graphite; (c) compacted (vermicular) graphite.

implemented on the molten iron just before pouring, graphite becomes round (b). Cast iron having this shape is called spheroidal graphite cast iron or nodular cast iron. The additive for spheroidizing is called a nodularizer. Compared to the flaky shape, this geometrical shape can avoid microstructural stress concentration to give higher mechanical strength and ductility. This is also referred to as ductile iron. However, the thermal conductivity and resistance to scuffing are not so high.

The third microstructure (c) is compact graphite iron (CGI) containing graphite of a vermicular (worm-like) form (c). This is a relatively new alloy that has improved mechanical strength without diminishing the favorable properties of flaky graphite iron. As may be inferred from the shape, the properties of this iron are positioned between flaky and spheroidal iron. It has a higher tensile strength being 1.5 to 2 times as strong as flaky iron, higher stiffness and approximately double the fatigue strength of flaky iron. The thermal conductivity lies between flaky and spheroidal iron. This makes it possible to produce a cylinder block that is both thinner and stronger.

The nodularizer inoculated in the molten iron gives perfect spheroidal graphite. An imperfect spheroidizing treatment before pouring generates CGI. The nodularizer contains Mg. After the inoculation of the nodularizer, the graphite shape gradually changes to a flaky shape via a vermicular shape as shown in Fig. 2.18.[6] The residual Mg content in the molten iron decreases with time due to the evaporative nature of Mg. This causes degradation of the spheroidal graphite to give CGI. An additional explanation is given in Chapter 4 and Appendix D.

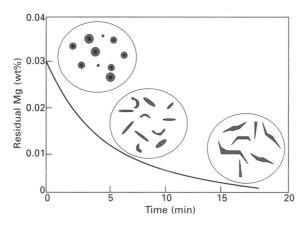

2.18 Effect of residual Mg on the graphite shape.

The number of cast iron blocks used for petrol engines has decreased, while CGI is seen as a new cast iron for diesel engines. To meet Euro IV (2005) emission regulations, cylinder pressures of 18 MPa and higher are being planned for car diesels. Even more stringent requirements under Euro V will come in from 2007 and 2008. Higher strength blocks with good thermal discharge properties will be required.

2.4   Aluminum blocks with enclosed cast iron liners to improve cooling performance

Until recently, with the exception of sports cars, car engines mainly used cast iron monolithic blocks. This was due to the fact that most cars did not require high power output and cast iron was inexpensive. Aluminum is weaker, than cast iron. It was thought that an aluminum block must be made much thicker than a cast iron block. However, if it is well designed, an aluminum block can be both much lighter and almost as strong as the cast iron block. In a recent comparison, an aluminum block can attain 40% reduction in weight compared to its cast iron equivalent. Despite the greater cost of aluminum, the demand for better fuel consumption has resulted in a significant increase in the production volume of light aluminum blocks. Aluminum blocks were used in 60% of European car engines in 2003.

Some basic designs of cylinder block are illustrated in Fig. 2.6. Figure 2.19[7] summarizes types of design for aluminum blocks, including modification technologies for the bore surface. The numbers in this figure correspond to those in Table 2.1. Referring to the figures and table, let us now focus on these characteristics. First, we will see blocks enclosing liners, followed by monolithic blocks.

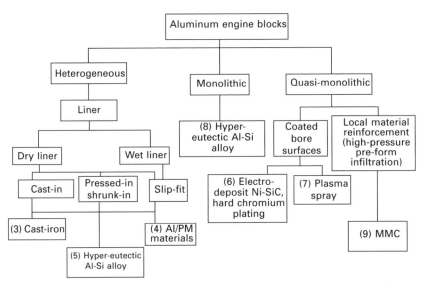

2.19 Aluminum engine block designs. The numbers correspond to the numbers in Table 2.1.

The aluminum alloy used for blocks is typically JIS-AC4B (Table 2.2), which has a thermal conductivity of 150 W/(m · K). Cast iron has a thermal conductivity as low as 50 W/(m · K). The thermal conductivity of aluminum is therefore three times that of cast iron. Since the density is 1/3 that of cast iron, aluminum alloy can give high cooling performance at a lower weight. However, it is soft and the wear resistance is generally low. To deal with this problem, aluminum alloy blocks with enclosed iron liners (normally cast iron) (3 in Table 2.1) are widely used. Historically, this composite structure was particularly developed for aeroplane engines which needed to be of light weight. For example, the Wright brother's engine (1903) powered the first self-propelled, piloted aircraft. It had a water-cooled, petrol, four-stroke, four-cylinder engine.[8] To lower the weight, a cast aluminum engine block containing a crankcase was used. The cylinder portion enclosed screwed steel liners.

In a block enclosing an iron liner, three types have been developed as shown in Fig. 2.6. The cast-in design is the most widespread. This is called a composite cast cylinder block. In this manufacturing process, molten aluminum is poured into the mold in which a cast iron liner is already placed. The hollow space surrounded by the liner is stuffed with a sand core (Fig. 2.43, described later). After casting, a two-metal structure is obtained in which the liner is firmly enclosed. This composite block has a residual shrink force on the liner. This is caused due to the fact that the larger thermal expansion coefficient of aluminum grips the cast iron liner during cooling after solidification. The thermal expansion coefficient of cast iron (10×10^{-6}/°C) is 1/2 that of aluminum (20×10^{-6}/°C). The stress remains as a residual stress after finish machining.

This stress works well when the engine temperature rises. Typically, the uppermost outside portion of the liner has a temperature of 250 °C in the air-cooling type, and 200 °C in the water-cooling type. The heat expands the aluminum portion more than the inside liner. Without the residual stress, the liner would loosen, generating a gap.[9]

The optimum microstructure suitable for the liner is the same as that required for the cast iron block (pearlite matrix containing flaky graphite). The cast iron liner is normally produced by sand casting. A centrifugal casting process using a rotating cylindrical die is sometimes used to produce a large bore liner. Since a liner is a thin tube, spray forming technology can also produce a steel liner.[10]

In this composite cast design, the iron liner does not come directly into contact with the coolant. The combustion heat must discharge to the coolant through aluminum. The heat should pass through the interface between the liner and aluminum. However, there can be some metallurgical discontinuity at the interface that may cause an air gap. The gap obstructs the heat transfer from the liner to the aluminum body. To reduce any gap, a number of techniques have been proposed. In permanent mold casting or sand casting, pre-heating the liner before cast-in is effective. The casting plan should be designed for the aluminum melt to flow along the surface of the liner without swirl. In high-pressure die casting, the pressurized liquid aluminum is injected at high speed, so that a gap is unlikely to appear.

To generate strong mechanical interlocking with aluminum, a cast iron liner having a dimpled outer surface has been proposed (Fig. 2.20). Figure 2.21 shows an example where the aluminum and liner join at the interface

2.20 Cast iron liner having dimpled outer surface.

without any visible gaps. The dimpled outer surface of the liner is caused by the coarse sand particles of the mold. The dimple gives an excellent heat transfer property with its large surface area and close contact. An alternative method has also been proposed. An aluminum coating on the outside surface of the cast iron liner creates good metallurgical continuity with the aluminum block. The coated layer works as a binding layer between the liner and block's aluminum. Dipping the liner directly into Al-Si alloy melt or thermal spray of Al-Si alloy[10] on the liner is used for coating. Even using this method it is, however, necessary to ensure careful casting.

The entire block will be prone to distortion over time. If the cast iron liner is thick enough, high rigidity prevents the liner from irregular distortion, irrespective of whether the liner has metallurgical continuity with the block

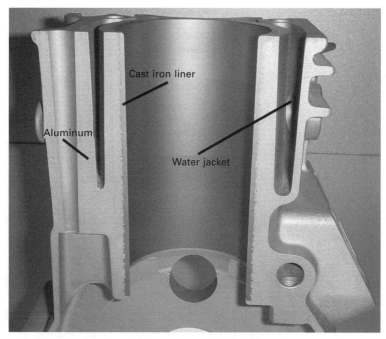

2.21 Cutaway of an aluminum block enclosing a cast iron liner having dimpled outer surface.

or not. Without metallurgical continuity, the liner can freely separate from the outer aluminum so that the bore shape is less affected by distortion in the block. However, in the case of a thin liner with less rigidity, distortion behavior is largely influenced by whether the liner is bonded metallurgically to the aluminum or held mechanically. With metallurgical continuity, the liner helps increase the rigidity of the cylinder and maintain the bore shape. In contrast, without metallurgical continuity, the liner can freely separate and does not help to increase the rigidity of the cylinder. The aluminum block should be carefully designed to compensate for this potential weakness.

2.5   Thermal distortion and heat discharge

2.5.1   How does the cylinder enclosing a press-fit liner deform with heat?

The liner is enclosed by either a press-fit or shrunk-in process (Fig. 2.6; 5 in Table 2.1). The press-fitted liner is a simple flanged tube (see Fig. 2.22). The liner should be perfectly finished before fitting. The finishing process includes grinding of the outside surface and honing of the inside bore wall. There is also a process where the honing of the inside bore wall is done after fitting.

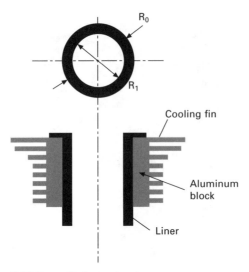

2.22 Press-fit liner structure.

If the liner is inserted into the cylinder body without further treatment, it unfastens as a result of the thermal expansion of aluminum. The correct design enables the bore to retain a perfectly round form when the liner elastically expands under the thermal expansion of the aluminum body. The external diameter of the liner shrinks from R_0 before fitting to R_1 after fitting. The interference ($= R_0 - R_1$) should be such that it does not cause buckling of the liner but, at the same time, provides a sufficient tightening force even at elevated temperatures. It is usually around 60 μm for a bore diameter of 60 mm.

The shrunk-in process is also used. It fastens the block and liner together by heating the outer block until it expands sufficiently to pass over the liner diameter and, on cooling, grips it tightly. This shrunk-in design is widespread in outboard marine engines. Where a cast iron block is used, the liner is inserted and held without interference because both materials are the same (2 in Table 2.1). Unlike the composite cast design, individual liners can be replaced once they become worn. Hence, replacing the entire block is avoided.

2.5.2 Powder metallurgical aluminum liner improves heat transfer

In a composite cast cylinder, the heat transfer through a cast iron liner is not that good. A thick liner reduces heat conduction, raising the bore wall temperature. A hard aluminum liner (4 in Table 2.1) has been proposed to deal with this problem. Honda has marketed a motorcycle engine using cylinder liners made from a rapidly solidified powder metallurgical (PM) aluminum alloy.[11, 12] The chemical composition of the liner is Al-17% Si-5 Fe-3.5 Cu-1 Mg-0.5 Mn containing Al_2O_3 and graphite. Figure 2.23 shows a typical microstructure. The hard Si particles (1200 HV) as well as finely dispersed intermetallic compounds embedded in the aluminum matrix give increased wear resistance. The liner is cast in by high-pressure die casting. The resulting wear resistance is nearly the same as that of cast iron. Daimler-Chrysler has also used a PM alloy cylinder liner[13] in its car engine. It is cast in by high-pressure die casting. This process is considered to be far more cost effective and it avoids the difficulties in tribological control of the hyper-eutectic Al-Si block (described later). The process used to achieve an optimal bore surface is almost the same as that used in producing a monolithic block of hyper-eutectic Al-Si alloy.

Figure 2.24 illustrates the production process. The molten alloy is sprayed and rapidly cooled into a powder first. The powder has a very fine microstructure during spraying. Next, the powder is canned in vacuum to make a billet for extrusion. Finally, the heated billet is hot-extruded into a tube. A spray forming process that directly deposits the sprayed powder to form a billet shape is also used. The powder particles weld together to form a bulk material during extrusion. The extruded tube is then cut into liners. The process generates the hardness and wear resistance needed for a liner.

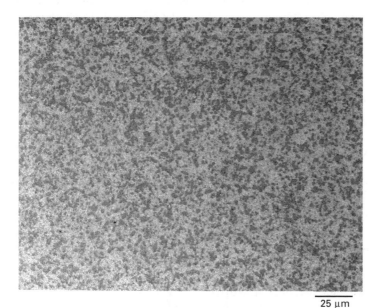

25 μm

2.23 Microstructure of powder metallurgical aluminum alloy.

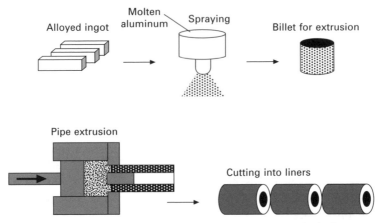

2.24 Manufacturing process of PM aluminum liner including spraying and extrusion.

2.6   Improving engine compaction with surface modifications

2.6.1   Shortening the bore interval

A low dimensional accuracy after machining or a thermal distortion during operation tends to cause poor clearance of the liner to the piston. Poor clearance removes necessary oil film, and at worst, can cause seizure. However, when the power output is not so high, the two-metal structure can lower the temperature sufficiently to maintain appropriate fluidity of the oil. The improved lubrication compensates for any poor clearance. This will be sufficient for an engine below about 100 HP a liter (74.5 kW/L) as a power output per unit displacement. However, at a higher power output range, this structure does not allow adequate cooling, due to the low thermal conductivity of the cast iron liner, and is likely to cause distortion. This problem can be addressed by redesigning the block.

Figure 2.25 illustrates the deck of a multi-cylinder block showing inter-bore spacing. The liner must have an appropriate wall thickness to keep its rigidity. If the liner does not have metallurgical continuity with aluminum, the required minimum thickness needs to be 2 mm. The jacket wall enclosing the liner also has to have enough thickness (a). As a general conclusion, these requirements cannot shorten the inter-bore spacing (b) to less than 8 mm.

2.6.2   Chromium plating

To make engines more compact, several linerless structures have been proposed.

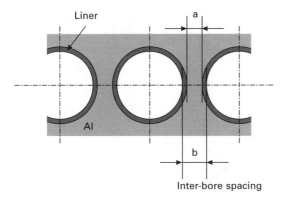

2.25 Inter-bore spacing of a cylinder block enclosing liners. Adequate thickness (a) of the aluminum wall is necessary to hold the liner. The decreased inter-bore spacing (b) makes the engine compact.

Monolithic and quasi-monolithic designs significantly decrease the bore wall temperature because of a higher heat transfer rate, generating less distortion. However, too narrow inter-bore spacing makes sealing by gasket difficult. A thin chromium layer directly plated onto an aluminum-alloy bore wall forms the running surface (6 in Table 2.1). The plated chromium shows good wear resistance because of its high hardness (800 HV). This process has been used since the invention of porous chromium plating by V Horst in 1942.[14] It generates finely dispersed cracks in the chromium layer, as shown in Fig. 2.26. The cracks retain oil to generate hydrodynamic lubrication. The technology was first used in the reciprocating engine of an aeroplane but more recently sports car and motorcycle manufacturers have also used this technology. The technology does not require special surface treatment at the piston surface.

A steel liner plated with chromium has also been proposed as a dry liner for diesel engines.[15] The plated chrome layer is, however, inferior in scuffing resistance (mentioned later). The disposal of waste fluid from the plating facility creates environmental problem. As an alternative, Ni-SiC composite plating is also used.

2.6.3   Ni-SiC composite plating

This process generates a Ni layer dispersing SiC particles[16, 17] (6 in Table 2.1). Figure 2.27 shows a water-cooled cylinder block coated with this plating. The piston surface does not require a special coating to prevent scuffing. Figure 2.28 is a cross-sectional microstructure of the plated layer. Figure 2.29 shows the magnified view of the Ni dispersing SiC particles. Polygonal SiC particles of about 2 μm can be seen. This plating was originally developed

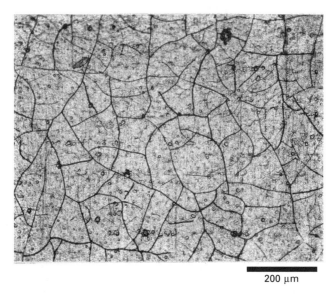

200 μm

2.26 Micro-cracks of plated Cr surface.

to coat the unique combustion chamber of a rotary engine in the 1960s. This plating forms a composite Ni layer containing particles or fibers.[18, 19] The SiC particle addition of around 4% is widely used. Cubic boron nitride (CBN) is also used,[20, 21] since its friction coefficient is lower than that of SiC.

2.27 Water-cooled cylinder block (closed deck type).

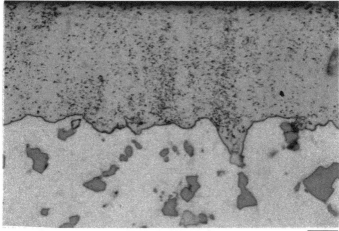

25 μm

2.28 Microstructure of Ni-SiC composite plating on the aluminum bore. The lower portion is aluminum.

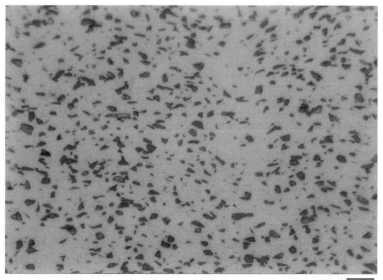

10 μm

2.29 Magnified view of dispersed SiC.

A small addition of phosphorus in the electrolyte[22] enriches P in the Ni layer, giving age hardening. Age hardening is an increase in hardness over time after exposure to elevated temperature, caused by small and uniformly dispersed precipitates (described in Chapter 3). Figure 2.30[23] compares hardness changes of some plated layers by heating (ageing); including Ni-SiC composite plating and hard chrome plating. The Ni-SiC plating without P (Ni-SiC) and the hard chrome plating continually soften with increasing temperature, while the Ni-SiC with P (Ni-P-SiC) increases in hardness up to 350 °C.

The resistance to scuffing of the plated chromium layer is lower than that of the Ni-SiC composite plating with P. The plated chromium contains chromium hydride just after plating. The hydride generates high lattice strain to raise the hardness of the chromium layer. As the hydride decomposes with heating, the chromium layer softens. This is why the hard chromium plating is not so resistive to scuffing. On the other hand, the Ni-SiC plating with P hardens with heat as shown in Fig. 2.30, which improves resistance to scuffing.

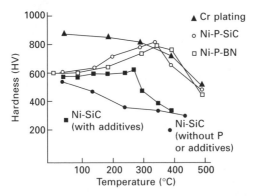

2.30 Hardness changes with heating. This is measured after holding for one hour at the indicated temperature. The P added specimens show higher hardness around 350 °C. The specimen with BN particulate (Ni-P-BN) is also shown.

A diamond whetstone finishes the composite plating to form an oil pocket. During operation, even if the Ni matrix wears, the SiC endures. Figure 2.31 schematically illustrates the changes of surface profile with wear. An advanced state of wear is shown on the right. The SiC particles need to have a size below 3 μm. If not, they disrupt the oil film, obstructing hydrodynamic lubrication. It is also very important to disperse the SiC particle homogeneously in the Ni layer. The electrolyte should be carefully stirred to homogeneously suspend the heavy SiC particles.

Plating is required only on the bore wall. However, in the conventional plating process, the entire block should be dipped into the electrolyte bath. Since the electrolyte corrodes aluminum, the portion not requiring plating must be protected. Since this is expensive, a process that plates only the bore wall has been developed.[24, 25] This technology allows electrolyte to flow only along the bore wall at high speed. A plating speed as fast as 10 to 40 μm/min has been accomplished[25] by raising the current density, while speeds of 0.8–3.0 μm/min are possible in conventional plating. Motorcycle engines with high power output have a particular requirement for high cooling

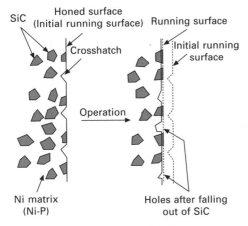

2.31 Schematic illustrations of the cross-sections of plating with Ni-SiC. Left, after honing; right, during operation. The SiC particle in the Ni matrix exposes during operation. The SiC particle buried in the plated Ni matrix crops out as the Ni wears.

performance at a light weight. The aluminum cylinder coated with Ni-SiC composite plating meets this requirement well.

2.6.4 Thermal spray

Kawasaki Heavy Industries produced a motorcycle cylinder coated with wire explosion spraying in 1973[26] (7 in Table 2.1). Initially this technology was used for two-stroke engines. The process thermal-sprays high-carbon steel and Mo alternately on the bore of an aluminum block (JIS-AC4B). The electrically heated wire melts and disperses to deposit on the bore surface. The Mo layer remains adhesing to the soft aluminum. Special coating is unnecessary at the piston surface. A plasma spray technique using metal powders has also been developed.[27, 28] It is very important for the sprayed layer to have close adhesion to aluminum, since it receives repetitive thermal stress during operation.

2.6.5 The hyper-eutectic Al-Si block

This technology makes a monolithic block with a hyper-eutectic Al-Si alloy (8 in Table 2.1), typically A390 (Table 2.2). The bore surface uses the aluminum matrix without coating to produce a wear resistant surface. The cast microstructure is shown in Fig. 2.32. Since this alloy is not easy to cast, low-pressure die casting (described later) has generally been used to obtain a sound casting. The Reynolds metal company originally proposed this technology.[29] GM adopted this for their Chevrolet-Vega model in 1971.[30, 31] After that, the technology spread to Germany. Yamaha has recently developed[32] the production of blocks by vacuum-assisted high-pressure die casting.

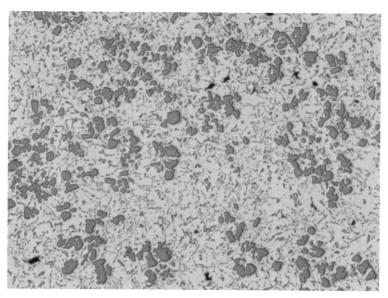

2.32 Microstructure of hyper-eutectic Al-Si.

The Si particles work in the same way as the SiC particles in Ni-SiC composite plating (Fig. 2.31). The Si dispersion in the casting must be carefully controlled from a tribological viewpoint. The hydrodynamic lubrication is greatest[32] at an appropriate height of the exposed Si. A special finishing to expose the Si is required for the running surface. To expose Si particles, the bore surface is chemically etched or mechanically finished after fine boring. Figure 2.33 is a scanning electron micrograph of the bore surface, showing primary phase Si particles of about 50 μm. Figure 2.34 is an atomic force micrograph, clearly showing exposed Si particles after finishing. The slightly depressed aluminum matrix between the Si particles helps retain oil. Figure 2.35 shows the three steps of the mechanical honing process.[33] The first pre-honing stage (a) adjusts the dimensional accuracy of the bore shape. Secondly, the basic honing (b) removes the broken Si. Finally, the finish-honing (c) recesses the aluminum matrix. The final surface roughness value measures around 1 to 3 μm in Rz value. Chemical etching is also used to expose Si particles.

Since the counter piston consists of a similar high-Si aluminum alloy, both running surfaces[34] become a combination of aluminum alloy. To avoid seizure, the piston surface is covered by plating such as Fe + Sn or Cr + Sn (the Sn layer being on the outside). The machining of the hard A390 alloy is not so easy. However, like a cast iron block, the fact that the block material forms the bore wall is attractive. As a result, production volumes are increasing.

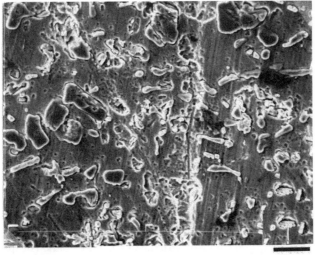

20 µm

2.33 Bore surface of an A390 alloy block. Si particles stand out from the matrix like stepping-stones.

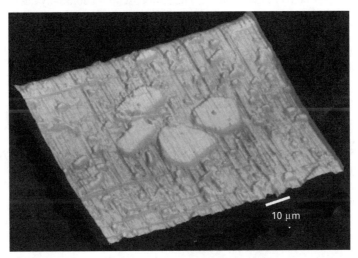

10 µm

2.34 Exposed Si crystals under atomic force microscopy.

2.6.6   Cast-in composite

Honda first installed this composite in its Prelude model (North America specification) in 1990[35] (9 in Table 2.1). A pre-form consisting of sapphire and carbon fibers is first set in the die. Then, medium-pressure die casting encloses the pre-form in the aluminum block. This process modifies the bore

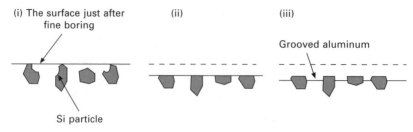

2.35 Three-step honing process to recess aluminum matrix. The broken line shows the position of the surface after fine boring.

wall into a composite material having high wear resistance.[36, 37] The surface becomes a metal matrix composite (MMC). The wear resistance is nearly the same as that of a cast iron liner. The piston should be plated by iron to prevent seizure. The average thickness of the block is greater and the production cycle time is longer compared to the standard high-pressure die casting block.[12] Similar technology has also been developed by other companies.[38] If the volume of fibers in the composite is high, the rigidity of the cylinder increases, reducing bore distortion. The Kolbenschmidt company has developed a similar technology forming a Si-rich-composite bore surface.[7] It casts the pre-form pipe consisting of a Si powder. Squeeze diecasting has aluminum melt penetrate into the porous pre-form. The MMCs in engines are listed in Appendix L.

2.7   Casting technologies for aluminum cylinder blocks

This section looks at the various casting technologies for aluminum. Typically, the mold material classifies casting technologies into either sand casting or die casting. Table 2.3 summarizes the technologies used. Table 2.4 lists the technologies for block casting and their characteristics.

Table 2.3 Casting technologies for aluminum blocks

	Gravity	Low pressure	High pressure ——————— Cold chamber	Squeeze
Sand mold	Cast iron, aluminum	Aluminum	Not applicable	Not applicable
Steel die	Aluminum	Aluminum	Aluminum, magnesium	Aluminum

Table 2.4 Mold materials for various casting processes

	Sand casting	Lost foam	High-pressure die casting		Gravity die casting (permanent mold casting)	Low-pressure die casting
			Conventional high-pressure die casting	Squeeze die casting		
Pressure (MPa)	Gravity	Gravity	100	70–150	Gravity	20
Dimensional accuracy	Low	Medium	High	High	Medium	Medium
Minimum thickness (mm)	3	3	2	4	3	3
Primary Si size in case of hyper-eutectic Al-Si (µm)	30–100	30–100	5–20	10–50	30–50	30–50
Gas content (cm³/100 g)	0.2–0.6	0.2–0.6	10–40	0.2–0.6	0.2–0.6	0.2–0.6
Quality — Blow hole	Medium	Medium	A lot	Few	Few	Few
Quality — Shrinkage defects	Less than a few	Less than a few	A lot at thick portion	Few	Less than a few	Less than a few
Quality — T6 heat treatment	Possible	Possible	Impossible	Possible	Possible	Possible
Quality — Welding	Possible	Possible	Impossible	Possible	Possible	Possible
Quality — Pressure tight	Low	Low	Good after resin impregnation	Excellent	Good	Good
Productivity*	100	80	100	50	50	40
Lifetime of the mold*	Mold pattern has long life	Mold pattern has long life	100	70	150	150
Cost*	150	160	100	130–170	150	200

*The ratio where conventional high-pressure die casting is 100.

2.7.1   Sand casting

Sand casting can produce aluminum blocks like those of cast iron. A typical resin bonded mold is shown in Fig. 2.36. In comparison with cast iron, sand casting of aluminum is not so easy because oxide entrapment and shrinkage defects are likely to occur. Cast iron does not cause this type of problem. Expansion during solidification due to the formation of graphite has made cast iron ideal for the economical production of shrinkage-free castings. Sand casting using resin bonded sand is normally used for aluminum.

The Cosworth process[39] is a low-pressure sand casting process used to obtain sound castings. An electro-magnetic pump fills molten aluminum from the bottom of the resin bonded mold. Large cylinder blocks have been produced using this method.

Core shaping water jacket

2.36 Resin bonded sand mold (partly disassembled) for an in-line four-cylinder block. The core shaping bores are removed and not shown.

Core package system[40] is a sand casting process proposed by Hydro Aluminum. A chemically bonded mold uses a bottom pouring plan and the mold is inverted after pouring. This inversion positions the runner portion at the top of the casting. Then the slowly solidifying runner portion feeds melt during solidification, which generates a porosity-free cylinder block.

2.7.2   Lost foam process

The lost foam process[41] uses a polystyrene foam having the same shape as that of the object to be cast. In normal sand casting, molten aluminum is poured into a cavity formed with bonded sand. In the lost foam process, the foam pattern made of polystyrene is embedded into unbonded sand. During pouring, the polystyrene foam pattern left in the sand is decomposed by molten metal. The casting traces the polystyrene shape.

The foam pattern must be produced for every casting made. This process starts with the pre-expansion of polystyrene beads, which contain pentane as a blowing agent. The pre-expanded beads are blown into a mold to form pattern sections. A steam cycle causes the beads to fully expand and fuse together in the mold. After cooling, the molded foam sections are assembled with glue, forming a cluster. The gating system (the passage liquid metal flows through) is also attached in a similar manner. Next the foam cluster is covered with a ceramic coating. The coating forms a barrier so that the molten metal does not penetrate or cause sand erosion during pouring. The coating also helps protect the structural integrity of the casting. After the coating dries, the cluster is placed into a flask and backed up with sand. Mold compaction is then achieved by using a vibration table to ensure uniform and proper compaction. Once this procedure is complete, the cluster is packed in the flask and the mold is ready to be poured. The molten metal replaces the foam, precisely duplicating all of the features of the pattern.

2.7.3   High-pressure die casting

In the die casting technique, the mold is generally not destroyed at each cast but is permanent, being made of die steel. Typically, three die casting processes are widespread: high-pressure die casting, gravity die casting and low-pressure die casting. In high-pressure die casting, liquid metal is injected into a steel die at high speed and high pressure (around 100 MPa). The process is illustrated in Fig. 2.37. This equipment consists of two vertical platens which hold the die halves. One platen is fixed and the other can move so that the die can be opened or closed.

A measured amount of molten metal is poured into the shot sleeve (a) and then introduced into the die cavity (b) using a hydraulically driven plunger. Once the metal has solidified, the die is opened (c) and the casting removed (d).

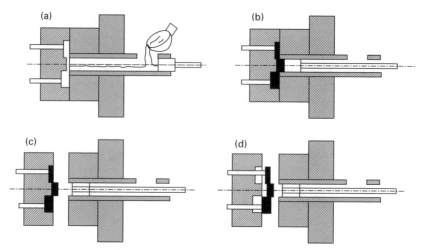

2.37 High-pressure die casting process: (a) the ladle cup doses molten aluminum into the injection sleeve after die closing; (b) the plunger injects the metal into the die cavity; (c) die parting after solidification; (d) the casting is ejected followed by die closing.

The die for a cylinder block enclosing a cast-in liner is illustrated in Fig. 2.38. The die temperature is kept relatively low compared to gravity and low-pressure die casting because the high injection speed enables the molten metal to fill the thin portion of a part without losing temperature. The low die

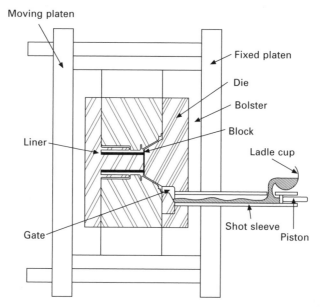

2.38 High-pressure die casting machine with die for an open deck block.

temperature (normally, around 200 °C) rapidly cools the part, so that this process gives not only a short cycle time but also good mechanical strength as cast. Other technologies require a thick protective coating sprayed on the die, which means looser tolerance and rougher surface finish. In high-pressure die casting, The low die temperature, as well as not needing a thick coating, gives a smooth surface and high dimensional accuracy (within 0.2% of the original cast dimensions). Hence, for many parts, post-machining can be totally eliminated, or very light machining may be required to bring dimensions to size.

From an economic aspect and for mass production of cylinder blocks, the short production cycle time rivals highly automated sand casting. Both the machine and its die are very expensive and for this reason high-pressure die casting is economical only for high-volume production. In this process, special precautions must be taken to avoid too much gas entrapment which causes blistering during subsequent heat treatment or welding of the product.

High-pressure die casting generally uses liquid aluminum, whilst semi-solid slurry is also used to get strong castings (Appendix J). The injected slurry including solid and liquid phases decreases shrinkage defects. Honda has recently introduced this process to produce a strong aluminum block[42] for a diesel engine.

2.7.4   Gravity die casting

In gravity die casting, like sand casting, the molten metal is gently poured into the cavity under gravitational force. The die temperature should be sufficiently high to ensure the molten metal fills all parts of the casting. The production cycle time is not always short because this process requires preparation of the steel die, and because the poured metal requires a long solidification time at high die temperature. The cost of the die is high but the process does not require expensive machinery like high-pressure die casting or a sophisticated sand disposal system like sand casting. This process is also called permanent mold casting.

2.7.5   Low-pressure die casting

In low-pressure die casting, the die is filled from a pressurized crucible below, and pressures of up to 20 MPa are usual. Figure 2.39 shows a schematic view of the process. In principle, low-pressure casting is possible for both sand molds and permanent molds as well as combinations. A crucible is positioned below the die, connected to the die through a vertical feed tube. When the chamber is pressurized, the metal level in the feed tube (fill stalk) goes up. The rising material slowly fills the die with little turbulence and without gas entrapment. The die temperature should be raised to get sufficient

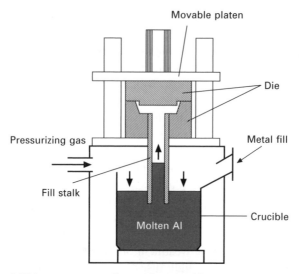

2.39 Low-pressure die casting machine.

metal filling. In comparison with high-pressure die casting, this process is suitable only for a medium walled casting. The hyper-eutectic Al-Si block normally uses this process to obtain a sound casting with fine dispersion of Si crystals (Table 2.4). Despite these benefits, the production cycle time is long.

2.7.6   Squeeze die casting

Squeeze die casting[43] is a high-pressure die casting process that can cast with the minimum of turbulence and gas entrapment. The use of very large gates and high hydraulic pressure makes the molten metal inject slowly into the cavity and pressurize just before solidification. This decreases shrinkage. In comparison with conventional high-pressure die casting, the result is a porosity-free, heat-treatable component with a thick wall capable of surviving the critical functional testing that is essential for structural automotive parts. This technique is mainly used for aluminum blocks having a bore wall of composite microstructure. The slow injection and medium-pressure squeeze assist aluminum infiltration. Additional casting processes for aluminum are explained in Appendix J.

2.8   Open and closed deck structures

The cylinder block deck is a flat, machined surface for the cylinder head. The head gasket and cylinder head fit onto the deck surface. For the head bolts, bolt holes are drilled and tapped into the deck. Coolant and oil passages allow fluids to circulate through the block, head gasket, and cylinder heads. Figure 2.40 shows a cross-section of a cylinder block with the corner removed, showing coolant passages. The water-cooling cylinder has water jackets allowing coolant to remove excess heat from the engine. There are two types. One is the open deck type (Fig. 2.41), and the other the closed deck type (Figs 2.2, 2.27 and 2.40). The difference is in the shape of the coolant passages at the mating plane to the cylinder head. The open deck type has a fully open coolant passage at the deck, whilst the closed deck type has a half-closed coolant passage. The closed deck type can increase the rigidity of the head gasket area to retain the roundness of the bore.

2.40 Cutaway of a water-cooled aluminum block.

The open deck type can be produced by all of the casting processes mentioned above but it is generally made by high-pressure die casting because the steel die can shape the water jacket as demonstrated in Fig. 2.38. Figure 2.21 shows a cross-section. By contrast, the coolant passage of the closed deck type has a half-closed shape as in Fig. 2.40. The passage widens within the block. Hence, the steel die cannot shape the water jacket. Instead, gravity die casting or sand mold casting using an expendable shell core (a resin bonded sand) is used.

In high-pressure die casting, a sand core is not generally used because the sand core is fragile and cannot endure the high-pressure injection. However, although there is some limitation in the shape, recent developments in sand core technologies have made it possible to produce a closed deck type using

2.41 Open deck block with integrated crankcase portion. The bores are plated with Ni-SiC composite.

high-pressure die casting.[44, 45] The use of these casting processes in the European market in 2003 was estimated to be: high-pressure die casting 56%, sand casting 19%, low-pressure die casting 11%, squeeze die casting 8% and lost foam process 6%.

2.9   The two-stroke cycle engine cylinder

Two-stroke petrol engines for motorcycles use a loop-scavenging method called the Schnuerrle system. Figure 2.42 shows a cutaway view of an air-cooled cylinder block. This is an aluminum block enclosing a cast-in liner. Unlike the simple cylinder shape of a four-stroke engine, the two-stroke cylinder has several portholes for inlet (suction), exhaust and scavenging in the bore surface. The portholes are connected to the gas passages in the block. The combustion gas flows in and out through these portholes.[46] Additionally, water-cooled engines have complex coolant passages. These shapes are cast as hollow shapes using a sand core.

Several types of cylinder structures have been developed. Low power output engines such as those for a small scooter often use cast iron monolithic blocks (1 in Table 2.1), because of their low cost. It is difficult to obtain a uniform pearlite microstructure along the cylinder bore because rapid solidification generates hard carbide at the thin rib between the portholes.[47] This is called chill (Appendix D and Chapter 5). When this appears, the hard microstructure lowers the dimensional accuracy of the bore after machining,

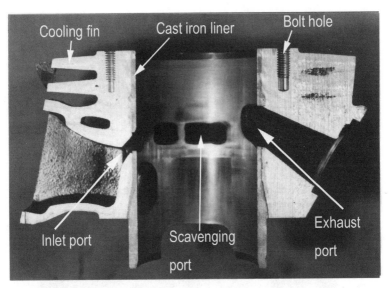

2.42 Crosscut view of an air-cooled two-stroke cylinder enclosing a cast iron liner.

damaging tribological properties. The high vanadium content (High-V cast iron) listed in Table 2.2 prevents chill and helps create a homogeneous pearlite microstructure.

The composite cast cylinder is used for models requiring high cooling performance (2 in Table 2.1). This two-metal structure encloses a cast iron liner having porthole openings. Figure 2.43 shows a shell core holding a cast iron liner. This integral core is placed into the holding mold. Molten aluminum is then poured into the mold. After shaking off the sand, a porthole opening connected to the gas passage like that in Fig. 2.42 is obtained.

High-performance models of two-stroke engines use surface modification methods like Ni-SiC composite plating (6 in Table 2.1). The area around the exhaust port is likely to overheat when the bore wall temperature is high, causing distortion of the bore wall.[48] A plated aluminum block is successfully used to avoid this unfavorable distortion and for better cooling performance.

Two-stroke outboard marine engines have comparatively simple porthole shapes. The cast iron liner is shrunk in to the aluminum block (5 in Table 2.1). It is difficult for the liner hole to accurately match up with the hole opening of the block.

2.10   Conclusions

Several solutions are possible to overcome problems in bore wall design. Various methods for forming hard layers on the aluminum surface have been discussed in this chapter and are summarized in reference.[49] In considering

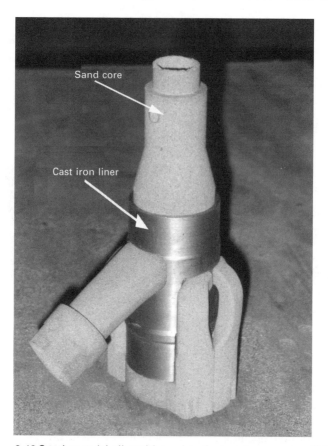

2.43 Sand core (shell mold core) to form gas passages. This pre-formed core with a liner is placed in the holding mold before pouring. The mold and core are decayed after casting, so that the aluminum block enclosing the cast-in liner remains. The complex gas (and coolant) circuits are three-dimensionally designed to have a proper flow. The core shapes hollow circuits in the block.

surface modification in engine cylinders, we should look not only at the surface, such as the hardness and the coefficient of friction, but also the quality of the substrate on which the modification is implemented. The castings must have a fine microstructure without holes.

Design technology in dealing with cooling, factors affecting distortion and machining accuracy are very important. The tribological matching between the piston and piston ring should also be taken into consideration. A high-output engine generates a lot of heat, frequently causing tribological problems. Appropriate cooling is necessary for efficient lubrication. In the case of aluminum cylinder blocks, the composite cast type enclosing a cast iron liner will be mainstream, while various surface modifications can be used to improve the performance of the engine.

2.11   References and notes

1. Maier K., *VDI Berichte*, 866 (1990) 99.
2. Hill S.H., *et al.*, SAE Paper 950938.
3. The contact pressure of the ring in two-stroke engines is relatively low in comparison with that of four-stroke engines. The lubrication oil mixed with the petrol is consumed regularly.
4. Yamamoto H., *et al.*, *Jidoushagijutuskai Kouen Maezurishu*, 934 (1993) 89 (in Japanese).
5. Matsuo S., *et al.*, *Proc. Int. Sympo. Automotive Technol. Autom.*, (1994) 183.
6. Huebler J. and Melnikova L., Private communication, (2003).
7. Koehler E., *et al.*, ATZ/MTZ Special Edition *Werkstoffe im Automobilbau*, (1996), 2.
8. Benson T., http://wright.nasa.gov/airplane/powered.html., (2003).
9. Tomituka K., *Nainenkikanno Rekishi*, Sanei Publishing, (1987) 85 (in Japanese).
10. Federal Mogul Corp., Catalogue, (2003).
11. Nikkei Materials & Technology, 142 (1994) 10 (in Japanese).
12. Koya E., *et al.*, Honda R & D *Technical Review*, 6 (1994) 126.
13. Kolbenschmidt Pierburg AG catalogue., http:// www.kolbenschmidt-pierburg.com, (2003).
14. Sekiyama S., *Kouku Daigakkou Kenkyuu Houkoku*, R32 (1980) 1 (in Japanese). For a motorcycle, the Kreidler company first used porous chrome plating on the aluminum cylinder of a two-stroke 50 cm^3 engine in 1950. Tomituka K. *Nainenkikanno Rekishi*, Tokyo, Sanei Publishing, (1987) 157 (in Japanese).
15. Tagami S., *Nainenkikan*, 29 (1990) 49 (in Japanese).
16. Maier K., *Oberflache*, 32 (1991) 18.
17. Funatani K., *et al.*, SAE Paper 940852.
18. Enomoto H., *et al.*, *Fukugoumekki*, Tokyo, Nikkan Kougyou Shinbunsha Publishing, (1986) (in Japanese).
19. Hayashi H., *Hyoumen Gijutsu*, 45 (1994) 1250 (in Japanese).
20. Konagai N., *et al.*, *Nainenkikan*, 33 (1994) 14 (in Japanese).
21. Muramatsu H. and Ishimori S., *Jidoushayou Arumi Hyoumenshori Kenkyukaishi*, 14 (1996) 17 (in Japanese).
22. Ishimori S., *et al.*, *Proceedings of the 71st. American Electroplater's Soc.*, (1984) 0–5.
23. Funatani S., *Kinzoku*, 65 (1995) 295 (in Japanese).
24. Emde V.W., *et al.*, *Galvanotech*, 86 (1995) 383.
25. Isobe M. and Ikegaya H., *Jidoushagijutu*, 48 (1994) 89 (in Japanese).
26. Fukunaga H., *et al.*, *Nihon Yousha Kyoukaishi*, 11 (1974) 193 (in Japanese).
27. Barbezat G. and Wuest G., *Surface engineering*, 14 (1998) 113.
28. Barbezat G., *et al.*, *Proceedings of the 1st united thermal spray conference, Indianapolis*, ASM international, (1997) 11.
29. Jorstad J.L., *Trans. American Foundrymen's Society*, 79 (1971) 85.
30. Jacobsen E.G., SAE Paper 830006.
31. Jorstad J.L. SAE Paper 830010.
32. Kurita H. and Yamagata H., SAE Paper 2004-01-1028.
33. Klink U. and Flores G., *Machinen Fabrik Gehring GmbH & Co.*, (2000).
34. In the small four-stroke engine, a high-pressure die cast cylinder bore is used without any surface modification and the piston is not coated, yet this engine generates only low power.

35. *Jidousha Kougaku*, 40 (1992) 38 (in Japanese).
36. Ebisawa M. and Hara T., SAE Paper 910835.
37. Shibata K. and Ushio H., *Tribology International*, 27 (1994) 39.
38. Takami T., *et al.*, SAE Paper 2000-01-1231.
39. Cosworth home page, http:// www.cosworth-technology.co.uk, (2003).
40. Hydro Aluminum homepage, http://www.hydro.com, (2005).
41. AFS home page, http://web.umr.edu/~foundry/first.htm, (2003).
42. Kuroki K., *et al.*, *Proceedings of the 8th International Conference on Semi-Solid Processing of Alloys and Composites, Limassol*, Paper No. 04-03, (2004).
43. Vinarcik E.J., *High Integrity Die Casting Processes*, New York, John Wiley & Sons, (2003) 51.
44. Mantani N. and Touhata T., *Sokeizai*, (1995) 14 (in Japanese).
45. Ryobi Co., Ltd., Catalogue, (2001) (in Japanese).
46. The port shape and number have changed to improve engine performance. Three ports for inlet, exhaust and scavenging were normal in the past, while seven ports are general now. Sufficient scavenging and air inlet efficiencies needed some additional ports. The exhaust port width has enlarged to decrease resistance to gas flow. These designs have made casting difficult.
47. The following ref. explains an invention on a pierced cast-iron liner having several ports. Koike T. and Yamagata H., *Advanced Technology of Plasticity* 1987, Springer-Verlag, (1987) 299.
48. Shirasagi S., *Nainenkikan*: 14 (1975) 11 (in Japanese).
49. *Aruminiumu Hyoumenno Atsumakukoukagijutsu*: JRCM, Tokyo, Nikkan Kougyou Shinbunsha Publishing, (1995) (in Japanese).

The piston

3.1   Structures and functions

3.1.1   Function

Figure 3.1 shows a typical assembly of pistons, connecting rods and a crankshaft for a four-stroke-cycle engine. The piston receives combustion power first

3.1 Forged pistons made of a continuously cast bar, connecting rods and a crankshaft for an in-line four-cylinder engine. The recess for the engine valve (5 valves) raises the compression ratio. Forging accurately stamps the recess shapes.

and then, the connecting rod transmits the power through the piston pin. Figure 3.2 shows pistons used for two-stroke engines. Figure 3.3 shows both inside and outside views of a four-stroke engine piston. Figure 3.4 illustrates an assembly comprising piston, piston ring, and piston pin.

3.2 Two-stroke engine pistons made of a powder metallurgical aluminum alloy. Two-stroke engines normally install two piston rings (Chapter 4). The long skirt is necessary for exchanging gas.

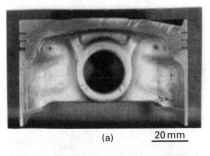

(a)   20 mm

(b)

3.3 Four-stroke engine piston, (a) inside and (b) outside. The skirt thickness is reduced to as small as 1.5 mm for light weight. Three piston-ring grooves are observable at both ends of the upper end. The piston-pin boss for the piston pin is positioned at the center. The high contact pressure exerted by the piston pin is likely to cause abrasive wear of the boss.

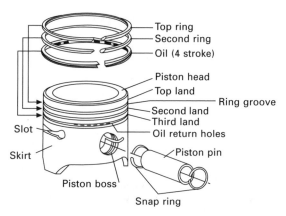

3.4 Nomenclature of each portion. The head and top-land areas reach the highest temperature because they directly contact with combustion gas. The piston ring placed into the ring groove is a spring sealing gas. The piston pin is a hollow tube made of carburized steel. The pin inserted in the pin boss is held at both ends with snap rings in order not to jump out. The piston pin revolves during operation.

A piston can best be described as a moving plug within the cylinder bore. Figure 3.5 summarizes its various functions. Firstly, the piston together with the piston rings, form the combustion chamber with the cylinder head, sealing the combustion gas. Second, it transfers combustion pressure to the rotating crankshaft through the piston pin and connecting rod. Third, in the two-stroke engine particularly, the piston itself works as the valve, exchanging gas (Chapter 2). Pistons also need to be mass-produced at low expense.

When exposed to a high-temperature gas, the piston must reciprocate at high speed along the cylinder bore. Even a small engine can generate a high power output at an increased revolution number. A petrol engine piston of 56 mm diameter can typically shoulder a load of about 20 kN and moves at a velocity as high as 15 m/s. If the number of revolutions is assumed to be 15,000 rpm, the piston material reaches the repetition number of 10^7 times in 11 hours. This number roughly equates to the fatigue-strength limit of a material. To generate high power output, the piston should be designed to be as light as possible whilst retaining durability. For example, the piston in Fig. 3.1 weighs 170 g whereas those in Fig. 3.2 are 150 g on average. Such lightweight designs increase the stress on the piston material.

In comparison with petrol engines, diesel engines generate power at relatively low rotation, although a higher cylinder pressure is required. Figure 3.6 shows a direct injection diesel piston with a combustion bowl. The increased demand on diesels today requires high performance at high efficiency with low emissions. Most of the direct injection diesel engines used in cars in the year 2003 reached specific power outputs of up to 40 kW/L. These engines

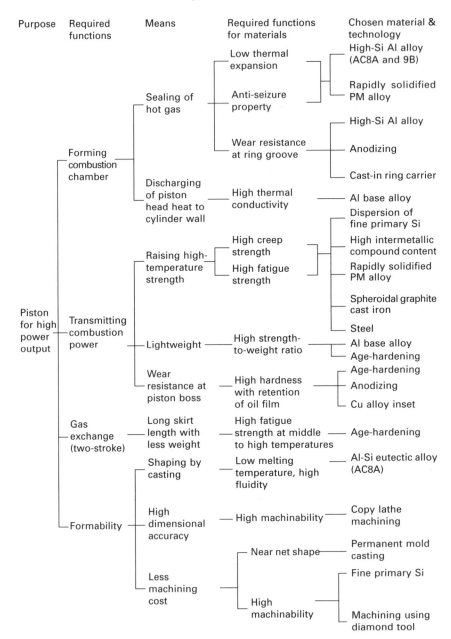

3.5 Functions of a piston for high output power.

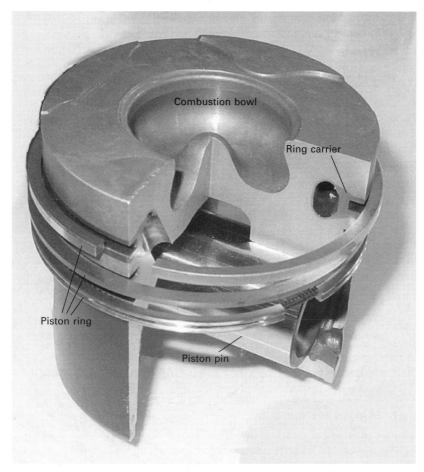

3.6 Aluminum piston for a direct injection diesel engine. The edge of the combustion bowl is fiber-reinforced. The top ring groove is reinforced with a Niresist ring carrier.

use a piston with a centrally located combustion bowl. The second generation of high-pressure injection systems (common rail and unit injector) has generated an increase in cylinder pressure up to peak values of 18 MPa. This increased power has raised the piston head temperature to as much as 350 °C at the combustion bowl edge. Compared to petrol engine pistons, diesels require a piston with a much higher strength to withstand these high temperatures.

Light aluminum alloy has been the most widely used material. The first aluminum-alloy piston appeared at the beginning of the 20th century just after the invention of the electrolytic smelting technology of aluminum in 1886. At that time, internal combustion engines used cast iron pistons. A key issue was whether aluminum, with a melting temperature as low as 660 °C, could withstand hot combustion gas. Figure 3.7 describes key requirements and reinforcement methods for aluminum and iron pistons.

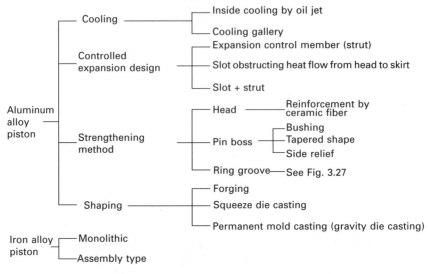

3.7 Structures and manufacturing processes of automotive pistons.

3.1.2   The use of Si to decrease the thermal expansion of aluminum

Figure 3.5 describes the various requirements of a piston. As the engine starts, the piston head heats up rapidly and expands in diameter. The cylinder block enclosing the pistons has a large heat capacity and employs a water-cooling system, so it heats up slowly. If the running clearance between the piston and the cylinder bore is too narrow, then the expanded piston will touch the bore wall instantly. If the clearance is too wide, it will cause blow-by of the combustion gas when the block warms up. When the engine bore is cold, a large running clearance will also make a noise because the piston knocks the bore wall. Hence, the piston material should have a low thermal expansion coefficient.

Pure aluminum has a thermal expansion coefficient up to 23.5×10^{-6} /°C. To decrease the thermal expansion coefficient, aluminum alloys containing Si as high as 12% or 19% are generally used. The alloyed Si reduces the thermal expansion coefficient because that of Si is as low as 9.6×10^{-6} /°C. A copper-alloyed aluminum (Y alloy; a heat-resisting Al-Cu-Mg alloy) was used until the Karlschmidt company began to use the 14% Si alloy (Si was added into the Y alloy) in 1924.[1] Si has additional advantages when alloyed in aluminum. It lowers the alloy's density and decreases the piston weight. Pure Al has a density of 2.67 g/cm^3, while pure Si 2.33 g/cm^3. Si also raises its wear resistance. The alloyed Si is also very effective in preventing seizure of the piston ring to the ring groove. Si has a Vicker's hardness value in the range of 870 to 1350 HV. Hard Si is a semi-metal, showing similar properties to those of ceramics.

Al-Si alloy thus has excellent characteristics as a piston material. Various improvements have been made to produce the alloy compositions shown in Table 3.1. JIS-AC8A is mainly used for four-stroke pistons and JIS-AC9B for two-stroke pistons. These alloys have a different Si content. A higher Si content beyond 12% decreases the thermal expansion even more. The AC8A has a thermal expansion coefficient of $21.0 \times 10^{-6}/°C$ and AC9B (19% Si) $19.9 \times 10^{-6}/°C$. Two-stroke engines generate a high power output, so the piston is likely to reach a much higher temperature. The piston reaches approximately 100 °C higher temperature than that of the four-stroke piston. This higher temperature is likely to cause seizure. The AC 9B alloy is generally used to increase the seizure resistance of the two-stroke engine piston.[2] The powder-metallurgical alloy AFP1 in Table 3.1 will be referred to later.

Table 3.1 Chemical compositions of piston alloys (%). JIS-AC8A and AC9B are for a cast piston. AFP1 is a rapidly solidified powder-metallurgical alloy for a forged piston

Piston alloy	Cu	Si	Mg	Fe	Ni	Al	SiC	Zr
JIS-AC8A	1	12	1	–	1	Balance	–	–
JIS-AC9B	1	19	1	–	1	Balance	–	–
AFP1	1	17	1	5.2	–	Balance	2	0.9

Cast iron pistons were mainly used for petrol engines in the past. The combination of cast iron cylinder and piston (Chapter 2) results in good sealability, because both parts have the same thermal expansion. The graphite in cast iron also provides good wear resistance as a solid lubricant.[3] However, cast iron is much too heavy for pistons operating at high revolution speeds. A heavy piston increases the load on the connecting rod and crankshaft so these parts need to be strengthened with thicker designs. Also, the low thermal conductivity of cast iron obstructs heat dissipation, so the resultant high piston temperature can cause the air-fuel mixture to ignite spontaneously. These are the main reasons why cast iron pistons are no longer used in petrol engines. Magnesium alloys are lighter than aluminum alloys, but a suitable low-thermal-expansion alloy has not yet been developed.

3.2   Manufacturing process

3.2.1   Casting

A modern piston has a very complex shape with a thin wall. At present, permanent mold casting (gravity die casting, Chapter 2) is the most popular manufacturing process.[4] Figure 3.8 illustrates the entire process starting from the initial casting. High Si aluminum alloy, particularly around 12% Si, has good castability. As indicated in the Al-Si phase diagram shown in Fig. 3.9, the melting temperature decreases when increasing the Si content up to 12.7%, after which the melting point increases. An alloy containing 12.7% Si has a melting temperature of 577 °C, which is lower than that of pure aluminum (660 °C). The eutectic composition (12.7% Si) solidifies immediately at 577 °C from the liquid state. The temperature range where liquid and solid phases co-exist, like iced water, does not exist at this composition. Therefore high fluidity is kept even at a low temperature, generating good castability. The AC8A alloy in Table 3.1 has such a composition.

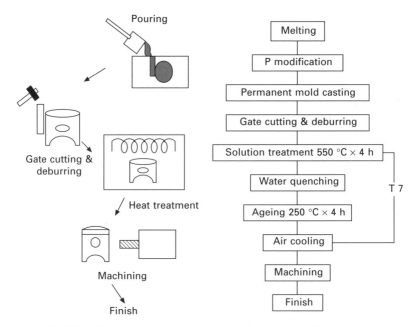

3.8 Manufacturing process of a cast piston made by permanent mold casting. T7 heat treatment includes four processes: solution treatment → water quenching → ageing → air cooling.

Figure 3.10 shows the typical microstructure of AC8A. Eutectic-Si crystals and intermetallic compounds are finely dispersed. The size as well as the distribution of these particles depends on the solidification rate; the faster the rate, the smaller the size and the finer the distribution. Figure 3.11 illustrates the solidification process from the melt phase.[5] The atoms move randomly in the melt (a). Upon cooling, crystal nuclei form in the melt (a → c). The nuclei grow in the liquid (c → d). Solidification terminates to give the polycrystalline state (f). Figure 3.12 shows the solidified microstructure of AC8A under low

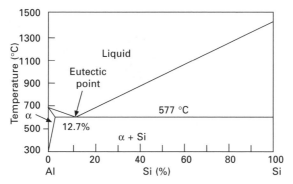

3.9 Al-Si phase diagram. The cast microstructure changes at 12.7% Si. Below 12.7% Si, α solid-solution crystals (the chemical composition is close to pure aluminum) appear first in the melt during solidification. Above 12.7% Si, coarse Si crystals called primary α appear first in the melt. It is called primary Si. By contrast, at 12.7% Si, the solidification takes place at once from the melt at 577 °C, without generating either primary Si or primary α. Hence, α and Si appear simultaneously. This type of solidification is called eutectic solidification. The chemical composition where the eutectic solidification takes place is called eutectic composition. The alloy having a Si content below the eutectic composition is called hypo-eutectic alloy. The alloy above the eutectic composition is called hyper-eutectic alloy. These defined terms are also used in the other alloy systems having eutectic solidification. The reader unfamiliar with phase diagrams could refer to Appendix C.

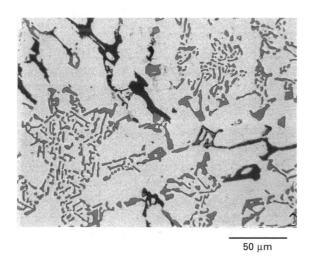

50 μm

3.10 Optical micrograph of AC8A. Gray particles are eutectic Si generated by eutectic solidification. The primary Si does not generate at this low Si content. Black skeleton-like particles are intermetallic compound $CuAl_2$.

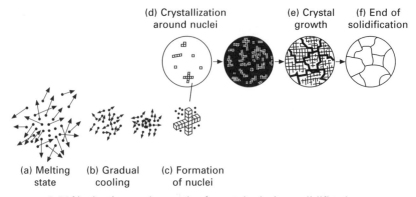

(d) Crystallization
around nuclei

(e) Crystal
growth

(f) End of
solidification

(a) Melting
state

(b) Gradual
cooling

(c) Formation
of nuclei

3.11 Nucleation and growth of crystals during solidification.

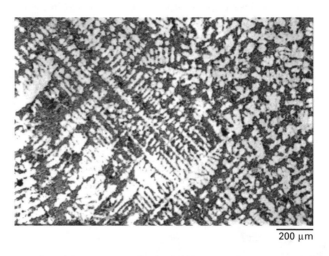

200 μm

3.12 Dendrite crystals of AC8A. White portions are crystals having a composition close to pure aluminum. Surrounding gray portions containing a higher Si concentration solidified at a later stage.

magnification. The white portions are crystals solidified as initial nuclei, having a composition close to that of pure aluminum. The white portions look like tree branches. These are called dendrite crystals. Figure 3.13 shows dendrite crystals observed in a shrinkage hole, a defect caused by improper casting. A high Si content above the eutectic point of 12.7% gives more favorable properties but casting becomes more difficult at these compositions because excess Si beyond 12.7% raises the melting temperature (Fig. 3.9).

3.2.2   Modifying the distribution of Si crystal

The second most important property of the piston listed in Fig. 3.5 is its ability to transmit combustion pressure. In order to do this, the piston material

100 μm

3.13 Dendrite crystals like tree branches observed in a shrinkage hole of AC8A.

requires high fatigue strength at high temperature. However, high-Si alloy is brittle because it inherits the brittleness of Si crystal. The brittleness needs to be overcome to raise fatigue strength. AC9B has the typical microstructure shown in Fig. 3.14. The Si in high-Si alloys appears as coarse crystals during the slow solidification of permanent mold casting. However, the addition of a small amount of phosphorus into the melt makes the Si crystals finer to reduce brittleness. Figure 3.15 shows the microstructure without the addition of phosphorus, and can be compared with the microstructure in Fig. 3.14 where phosphorus has been added. The effect is quite remarkable.

Prior to casting, the phosphorus ranging from 10 to 100 ppm is inoculated into the melt. This small amount of P forms AlP compound in the melt. AlP works as a solidification nucleus to make fine primary Si as small as 50 μm. With the increased P content, the primary Si size drastically decreases as shown in Fig. 3.16,[6] which in turn raises strength thus reducing brittleness. For AC8A alloy, instead of P, a small amount of Na or Sr is added to make the eutectic Si crystal fine. Figure 3.10 shows a micrograph after the addition of sodium. Because these alloys are so hard, diamond cutting tools[7] are used for machining and ring groove turning. The heat treatment that is used after casting (Fig. 3.8) will be described later.

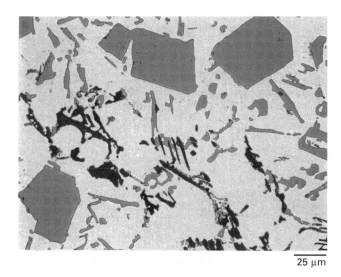

25 μm

3.14 Microstructure of AC9B alloy with phosphorus modification. The primary Si size measures around 50 μm. Small gray crystals are eutectic Si. Black skeleton-like crystals are intermetallic compound $CuAl_2$.

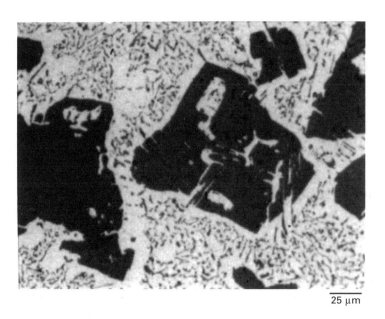

25 μm

3.15 Microstructure of AC9B alloy without phosphorus modification. Giant primary Si crystals are observable.

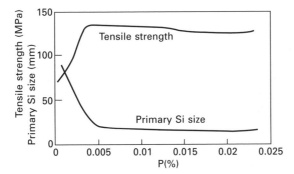

3.16 Influence of phosphorus on primary Si size and tensile strength. The base alloy is Al-22% Si.

3.3   Piston design to compensate thermal expansion

Increased Si in aluminum helps deal with the large thermal expansion coefficient of aluminum. However, an appropriate design can also help resolve this problem. The ideal design has a piston with a perfect round shape moving in a cylinder also with a perfect round shape at an appropriate running clearance under hydrodynamic lubrication. However, this is difficult in part because a piston has a complicated temperature distribution from its head to skirt. The temperature at the piston head is high (about 350 °C), whilst that of the skirt is low (about 150 °C). The difference in temperatures is likely to cause distortion during operation. Additionally, the gas pressure, inertia force and lateral force also increase the distortion. It is therefore very difficult to keep a constant running clearance.

Figure 3.17 illustrates the piston shape, also referred to as the piston curve. The piston head has a smaller diameter than that of the skirt, so a homogeneous running clearance to the cylinder wall should be maintained during operation. If the temperatures are assumed to be 350 °C at the head and 150 °C at the skirt, there is a temperature difference of 200 °C from top to bottom. AC9B has a thermal expansion coefficient of $19.9 \times 10^{-6}/°C$. The AC9B piston with a diameter of 56 mm shows an expansion value of 223 μm ($= 200 °C \times 56 \times 10^3 \times 19.9 \times 10^{-6}/°C$). Therefore the head diameter after machining should be 223 μm smaller than the skirt diameter.

A distorted elliptical form is sometimes used to accommodate distortion. Figure 3.18 indicates the out-of-roundness at the piston center of a two-stroke piston. The short axis of the ellipse is along the pin boss direction. This shape allows for the elastic deformation during operation in advance. The roundness mismatch after machining is measured by a roundness-measuring device. The combustion pressure elastically expands the short axis length and shortens the long axis length. Figure 3.19 shows the effect of

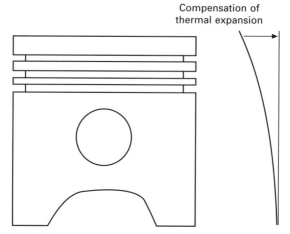

3.17 Piston curve having a slight spread from head to skirt. The diameter of the piston head expands as the arrow at elevated temperature.

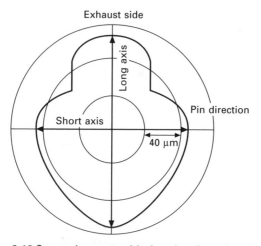

3.18 Curve shape considering the distortion during operation. The distorted ellipse has a short axis along the horizontal direction (pin direction).

applied pressure (indicated by load) on the length of the long axis. A contraction of about 50 μm is observable at the maximum pressure.

A standard piston design thus has a taper form along the vertical axis together with an elliptical cross-section. Only experimental analysis[8] and calculation[9, 10] can determine the proper shape. In order to have good reproducibility, mass-production machining is carried out using a copy lathe. Figure 3.20 describes the principle of the copy lathe for final machining,

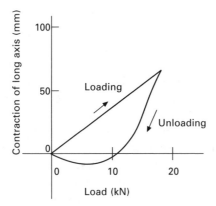

3.19 Elastic deformation of a piston skirt under load, measured in the long axis direction. The hysteresis is due to the friction between the piston-boss and piston-pin.

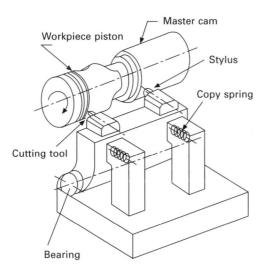

3.20 Copy lathe using a master cam (model). The stylus follows the shape of the master cam. The cutting tool fixed at the toolpost is forced to the workpiece by the spring. The bit (cutting tool) moves back and forth tracing the motion of the stylus. An alternative method is to use a numerically controlled lathe without a master cam.

which gives an accuracy of within several μm. It is, however, worth mentioning that the delicate piston shape results from the accumulated know-how of each individual manufacturer.

The final piston design must also allow for the head swing motion revolving around the piston pin. It is important to consider not only short-term deformation but also long-term creep deformation. Creep deformation is a phenomenon when material exposed to elevated temperatures gradually deforms even at a low stress.

In aluminum alloy pistons with cast iron blocks, other technologies have also been used to control thermal expansion, (listed in Fig. 3.7). The running clearance between piston and bore wall should be small enough to decrease noise, however, the clearance value must be designed for the piston skirt not to touch the bore wall after thermal expansion. This means that if the skirt does not expand or cause an increase in temperature, then it is possible to decrease the clearance. Two types of technologies have been proposed for this purpose. One is to make a slot as shown in Fig. 3.4, so that it keeps the skirt's temperature low by intercepting the heat flow from head to skirt. This slot is placed just below the oil ring groove. The disadvantage is that the head is likely to be overheated because the slot obstructs the heat discharge to the skirt, and the slot also lowers the strength of the piston. The other method is a controlled-expansion piston that includes small cast-in steel plates around the pin boss area. The cast-in steel plate mechanically restricts thermal expansion, so that the clearance to the cylinder bore does not decrease when the piston temperature is raised. In the past, this type of method was generally widespread, but the use of aluminum blocks has decreased its use.

3.4   Heat treatment

3.4.1   Age hardening and age softening

To increase its strength, a heat treatment called T7 is carried out on the cast piston. (Fig. 3.8). T7 is an age-hardening heat treatment typical to aluminum alloys. Figure 3.21 shows the hardness change of Al-4% Cu alloy as a function of time.[11] The alloy is water-quenched from 500 °C and then kept at 190 °C for the period indicated in the horizontal axis. The water-quenched Al-Cu alloy hardens in 3 to 30 hours (age hardening), and gradually softens over time (age softening).

Figure 3.22 shows the Al-Cu phase diagram. The water quenching is carried out along the broken line. The ageing process is schematically illustrated with scales in Fig. 3.23.[12] In Fig. 3.23(a), Cu atoms disperse randomly in the aluminum matrix after quenching. This state is called α-solid solution. In Fig. 3.23(b), the ageing at 190 °C aggregates Cu atoms. This state is called coherent precipitation of Cu, which increases lattice strain and produces hardening. This atomic movement takes place during ageing. In Fig. 3.23(c), the additional heating over a long period of time changes the coherent precipitation into the stable $\alpha + \theta$. The coherency between the precipitate lattice and matrix lattice observed in (b) disappears at this state. Consequently, the hardness and strength decreases. This is called age softening, being observable beyond 30 hours in Fig. 3.21.

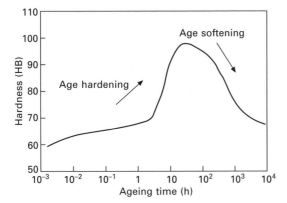

3.21 Hardness change of Al-4% Cu alloy at 190 °C. The vertical axis is indicated by Brinell hardness (HB). Hardening followed by softening is observable with time. The stage up to the peak is called age hardening. After the peak, the hardness decreases over a long period of time. It is called over-ageing or age softening.

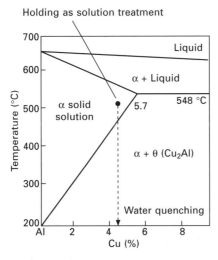

3.22 Al-Cu phase diagram. Cu crystallizes and disperses heterogeneously during casting. It dissolves homogeneously into the matrix by remaining at 500 °C. The following quenching stabilizes this state. This process is called solution treatment. Age hardening requires this solution treatment as a first step. Slow cooling instead of quenching from 500 °C does not raise hardness after ageing. The slow cooling generates $\alpha + \theta$ phase, where θ is intermetallic compound Cu_2Al. This precipitation obstructs age hardening.

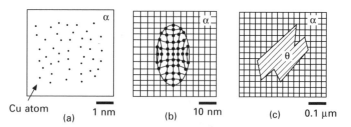

Cu atom   (a)   1 nm   (b)   10 nm   (c)   0.1 μm

3.23 Schematic illustration of age hardening process. The scale roughly indicates the size of the precipitation. (a) α-solid solution. Cu atoms disperse randomly among Al atoms. (b) Cu atoms precipitate coherently in α-solid solution. The grid schematically illustrates a crystal lattice. The term coherency means that the lattice plane of the precipitate matches that of the α-solid solution. (c) θ grows incoherently in the α-solid solution. The incoherency means that the crystal lattices of both phases do not match each other. The coherent precipitation (b) generates large lattice distortion, so that age hardening takes place only at this state.

The ageing mechanism has been explained using the simple Al-Cu system. The 1% Cu and 1% Mg in both AC 8 A and AC 9B strengthen the Al matrix by ageing. In these piston alloys, hardening takes place at the intermediate stage prior to the precipitation of the stable θ or S (intermetallic compound Al_5CuMg_2) phase. In general, ageing is used frequently to harden aluminum alloys. T7 heat treatment is widespread for pistons (Fig. 3.8).

3.4.2   Hardness measurement estimates the piston temperature during operation[13]

Pistons are used in the hardened state after the application of T7 heat treatment. However, a piston operated above 150 °C gradually causes age softening in the long term, so that the attained hardness obtained after T7 is not kept constant during engine operation. Figure 3.24 shows the decrease in hardness at three different temperatures. The material is AC8A hardened to 97 HRF (Rockwell hardness scale F) by T7 heat treatment. For example, after one hour, the hardness values measure 74 at 320 °C, 82 at 270 °C and 97 at 200 °C. It is generally accepted that the higher the holding temperature, the faster the softening. This is because atomic diffusion controls softening. Figure 3.25 shows the hardness values of a four-stroke piston after operation for 1 hour at 8500 rpm. The piston has a diameter of 67 mm. The lower part below the boss is omitted. The observed data is indicated in two cross-sections. One is on the inlet (IN)-exhaust (EX) direction plane, and the other is on the left (L)-right (R) direction. Both directions are at a right-angle. The minimum hardness appears at the piston center. If the hardness decrease is large, then the piston cannot withstand operation.

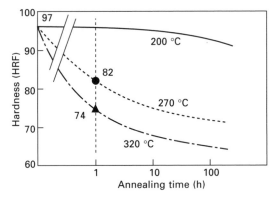

3.24 Hardness decrease of AC8A with heating (annealing).

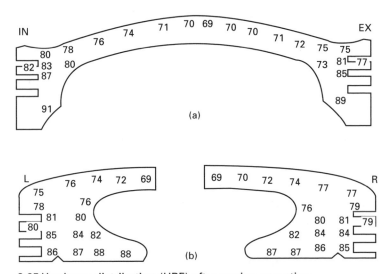

3.25 Hardness distribution (HRF) after engine operation.

This hardness decrease data can be used in developing a high-performance engine. It is very important to estimate the piston temperature during operation. The piston temperatures during operation can be measured by observing age-softening behavior, as the following example illustrates. To begin with, the entire piston body has a homogeneous hardness of 97 after T7 treatment. After engine operation, if the hardness values at two different points in the piston inside measure 82 (marked ● in Fig. 3.24) and 74 (▲), then the temperatures at each point are estimated to be 270 °C and 320 °C, respectively. From master curves like those in Fig. 3.24, it is possible to deduce temperatures after a fixed operation time.[14] Figure 3.26 illustrates the temperatures corresponding to Fig. 3.25. A higher temperature can be seen at the exhaust side.

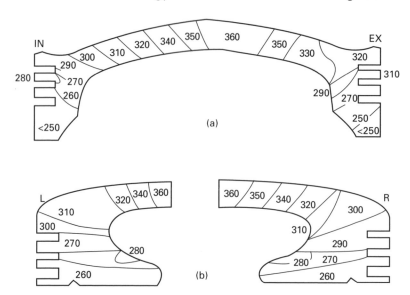

3.26 Estimated temperatures (°C) at two cross-sections, (a) IN (inlet valve)-EX (exhaust valve) direction and (b) L (left)-R (right) direction.

Some alternative methods have also been proposed. One of them is to measure the temperature with a thermocouple. Installing a fine thermocouple with a good response in the piston can measure the piston temperature even at a transient state. However, in order to equip the thermocouple whilst moving at high speeds, a link mechanism to wire the piston is required.[15] This installation is troublesome, so measurement using age softening is the preferred method.

3.5   Reinforcement of the piston ring groove

Abnormal wear at the ring groove not only decreases output power, but also increases oil consumption. At worst, seizure can take place. Groove wear is strongly influenced by piston thermal deformation under sustained working conditions. Ideally, the lateral faces of both the ring and the groove should be parallel and perpendicular to the cylinder bore wall, so that the contact and the resulting pressures can be distributed evenly, minimizing wear. However, the thermal distortion generates an inclination in the groove[16] to give localized contact between the lateral faces of the groove and the ring. Of course, too high a temperature in the ring groove softens aluminum which accelerates wear, but much wear is the result of distortion.

Several methods listed in Fig. 3.27[8] have been proposed to reinforce the piston-ring groove. During piston operation, the area particularly around the top-ring groove is liable to overheat. The aluminum softens and sometimes

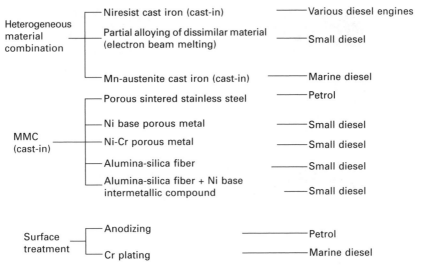

3.27 Reinforcement of piston-ring grooves. Except for anodizing, these are mainly used for diesel engine pistons. MMC stands for metal matrix composite.

fuses with the top ring. This fusing is also termed as micro-welding. Hard anodizing (hard anodic oxide coating)[17] is often used to avoid this fusing. Anodizing is a surface treatment widely used (refer to Appendix H) for aluminum appliances. When a piece of aluminum is anodized, a thin oxide layer is formed on its surface. Hard anodizing produces a particularly hard layer. To obtain a suitable hardness, a piston is anodized in a dilute sulfuric acid electrolyte at a low temperature. Figure 3.28 shows cross-sectional micrographs of anodized layers. Pure aluminum generates a homogeneous anodized layer (a). The layer of the piston alloy contains Si particles and intermetallic compounds which come from the substrate alloy (b).

Figure 3.29 shows the anodizing process over time. An aluminum surface is naturally covered with a thin oxide film. The dissociation in the sulfuric acid creates a thick oxide layer containing long pores. The film grows by changing aluminum into oxide. This tranformation accompanies expansion. In the film thickness, 2/3 is generated below the original substrate surface and 1/3 is formed above it (shown in 5). The oxide layer is nonconductive, which might inhibit thickening. However, the included pores in the oxide layer let ions into the aluminum substrate, which makes the oxide thick. Since the pore can accelerate corrosion, it can be closed by a steam treatment. The surface is used as a protective coating. The pore can retain various additives on the surface such as a lubricating oil. To raise wear resistance, a solid lubricant MoS_2 is intentionally filled in the pore.

Diesel engine pistons have a head temperature 50 to 100 °C higher than

(a)

(b)   25 μm

3.28 (a) Cross-sectional view of the anodized layer on pure aluminum and (b) on AC8A. The anodized layer (b) includes Si and intermetallic compounds. Anodizing is used not only for the piston-ring groove but for the piston head, as a protective coating preventing the heat-fatigue failure of a head. It also prevents fretting wear of the piston boss.

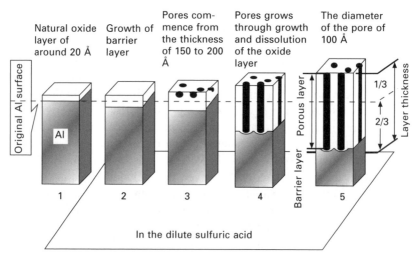

3.29 Growth of anodic oxide layer. Anodizing proceeds with time from left to right.

that of petrol engines. The high cylinder pressure exerts increased loading to the piston ring which in turn generates high contact pressure to the ring groove. Under such severe conditions, good durability is required. The wear resistance of anodizing is not sufficient at these very high temperatures, so the ring groove of diesel engines is cooled by an oil gallery and partially reinforced as shown in Fig. 3.6. The reinforcement called a ring carrier or a ring insert is buried to support and strengthen the piston ring groove. Various proposed technologies are listed in Fig. 3.27.

Figure 3.30 illustrates the reinforcement member method. An annular ring carrier (a) typically of Niresist cast iron is frequently cast into and positioned at the top-ring groove area.[8] Figure 3.30(c) indicates the area enclosing the carrier. Niresist comes from the group of cast irons containing a high amount of Ni. It has an austenite matrix dispersing graphite and Cr carbide. A typical chemical composition is: Fe-2.7% C-2.5 Si-1.2 Mn-17 Ni-2 Cr-6 Cu. Ni is an element that austenitizes iron. Niresist has a thermal expansion coefficient ($17.5 \times 10^{-6}/°C$) nearly the same as that of the piston alloy, because of its austenite crystal structure. This avoids unwanted distortion at the interface between the insert and aluminum. The hardness is as high as 170 HB.

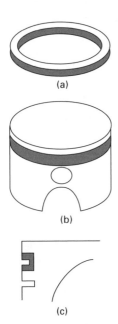

(a)

(b)

(c)

3.30 Piston having a cast-in ring carrier. (a) The insert (ring carrier) before casting. (b) Cast-in. (c) Cross cut view of the land area.

Since Niresist is an iron, it is heavy. As a lighter alternative, a pre-form of a mixture of alumina and silica fibers (the shape is much the same as in Fig. 3.30(a)) is also used for reinforcement.[18] This is cast-in by the squeeze die casting method (Chapter 2). The pre-form looks something like candyfloss. The aluminum melt forcibly infiltrates into the pre-form by squeeze die casting. The solidified portion becomes a metal matrix composite (MMC). This method is also used to reinforce the edge of the combustion bowl of diesel pistons and the entire piston head of petrol engine pistons.[19]

Partial remelting of the ring groove area can also increase hardness due to the fine microstructure generated by rapid solidification. In addition to this, if alloying is implemented during remelting, the groove is strengthened. A dissimilar metal such as Cu is added in the melt using electron beam melting.

The cast-in ring insert should bond with the piston alloy for good thermal conduction. Niresist is, however, a dissimilar metal from the piston alloy. It is impossible to get acceptable metallurgical bonding using just the cast-in procedure. For this reason, the insert is coated with aluminum prior to the cast-in. The insert is immersed and heated in the melt aluminum. The coated layer works as a binder during the cast-in procedure to give good metallurgical bonding with the piston body (Fig. 3.6 and Fig. 3.30(c)).

3.6   The high-strength piston

3.6.1   Strength of piston materials at high temperatures

Figure 3.31 shows an example of a fatigue crack observed at the piston head. A unique ring pattern is observable. This is called a beach mark, which typically appears in fatigue failure. When such a failure appears, the combustion conditions are adjusted or the piston head is made thicker to increase its strength. The current requirements of high power to weight ratio, and the

3.31 Fatigue fracture at a piston head. A high combustion pressure generated a high bending stress at the piston head.

trend to raise piston temperatures even more, demand an aluminum alloy with high strength at elevated temperatures. Age hardening strengthens piston alloys and the strength is kept at lower temperature parts of the piston, but it disappears in higher temperature areas, particularly at the center of the head. Age hardening cannot guarantee strength in these conditions. Figure 3.32 shows the variation of the strength of AC8A and AC9B at high temperatures. You can see their strength decreases rapidly above 250 °C.[20]

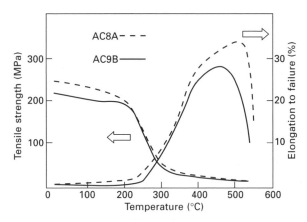

3.32 Temperature dependence of the tensile strength of AC8A and AC9B. Over-aged specimens were used to remove the effect of age hardening.

In the temperature range where the age hardening effect disappears, the base strength of the alloy becomes important. It is generally known that strength at high temperatures depends on the distribution as well as the quantity of dispersed intermetallic compounds (Appendix G). For example, alloyed Ni forms an intermetallic compound Al_3Ni during solidification, which contributes to high temperature strength. Generally, the higher the quantity of intermetallic compounds, the higher the strength. However, it is difficult to increase the content above a certain limited amount. The increased content coarsens the intermetallic compound, which in turn makes the material brittle (the reasons are mentioned in the next section).

The piston head temperature of diesels reaches as high as 350 °C. As has been mentioned, above 250 °C, the strength of aluminum alloy drastically decreases. In recent direct injection diesels, the high temperature area appears particularly at the combustion bowl. To raise the strength of the bowl edge, a ceramic fiber can be partially cast in to form a composite-like ring carrier.

3.6.2   The lightweight forged piston

There are particular design issues when manufacturing lightweight pistons. Figure 3.33 shows a lightweight-design construction diagram[21] with the so-

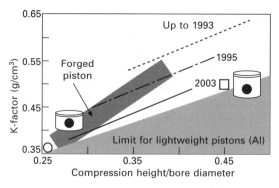

3.33 Lightweight design. Forged piston and cast piston shown in Fig. 3.34 located at positions marked ○ and □, respectively.

called apparent density of the piston (K factor), plotted vs. the relative compression height. The K factor is calculated by the equation $K = m/d^3$, where m is the weight of the piston and d the diameter. The relative compression height is defined as the compression height over the diameter of the cylinder. By decreasing the height, the piston shape changes from that of a deep cup to a shallow dish as illustrated in the figure. In the direction towards low compression height, the piston design is limited by the space needed for the ring zone and the connecting rod clearance as well as increasing piston head stresses.

If pistons are designed with the same principle, there is virtually a linear relationship between the K factor and relative compression height. Older designs favoured higher apparent density, but there has been a more recent emphasis on a more lightweight design. Figure 3.34 compares a cast piston for a car engine with that of a forged piston for an off-road motorcycle. The

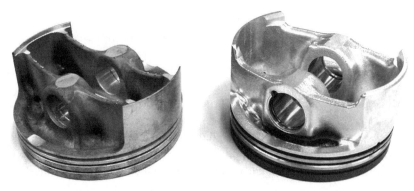

3.34 Forged piston (left) and cast piston (right). The left weighs 255g, the compression height is 23 mm and total height is 41 mm. The right weighs 360g, the compression height is 28 mm and total height is 50 mm.

diameters of the pistons are nearly the same (90 mm diameter), but the weights are different. At a constant compression height, the reduction in weight is constrained by the minimum wall thickness that can be achieved with aluminum and the wall thicknesses required to absorb lateral piston forces. The piston with decreased compression height enables a lightweight design, but it inevitably generates high bending stress at its head. The stress causes fatigue fracture as shown in Fig. 3.31. This means that the raised fatigue strength of a material can decrease the relative compression height to enable a much lighter-weight design.

One solution that breaks the limit is a forged piston. Casting can form complicated piston shapes. However, despite recent improvements, casting has some limitations in raising fatigue strength. For example, even after the degassing treatment of aluminum melt, gas defects such as blowholes cannot perfectly be eliminated. Thick parts of the casting have low fatigue strength because of the coarse microstructure generated by the slow solidification rate. An aluminum alloy suitable for a forged piston should contain Si. This alloy shows very small elongation as low as 2% (Fig. 3.33) and has low malleability at room temperature. Malleability is improved at elevated temperatures as shown in Fig. 3.32. The elongation value measures around 30% at 450 °C. This makes forging at elevated temperatures, often referred to as hot forging, possible. A more detailed description of forging technologies is given in Chapter 8. As shown in Fig. 3.32, high-Si aluminum is soft with maximum elongation around 450 °C, but the elongation is 30% at most. Hence, low malleability still makes forging difficult, particularly when forming complex shapes. To enable a more intricate shape, the shaped material should have an increased wall thickness and must not be of a near-net-shape. This is one of the reasons why the forged piston is expensive in comparison to that of a cast piston. Despite the fact that forging can generate a strong piston, high cost limits its use to very specialist engines.

In conventional hot forging of aluminum, the die temperature is not accurately controlled. The forger pre-heats the die with a burner to give the desired temperature. The billet for forging is also heated. These temperatures are determined experimentally. In actual production, the die temperature reaches a constant value depending on the cycle time of the forging. It becomes low at long cycle times. The die absorbs heat from the thin portions of the shaped material to cause cracking. In order not to lose heat, the shaped material inevitably becomes thicker and therefore needs additional machining to obtain a thin walled shape. Otherwise, additional forging and reheating is required. These extra procedures raise the cost. To produce a precise forged piston at a low cost, a technology called controlled forging[22, 23] has been developed. Figure 3.35 illustrates the process.

1.   The installed heater controls the forging temperature of the material, enabling the shaping of the thin portion.

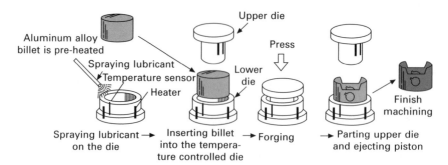

3.35 Controlled forging process. Generally, metal softens and ductility increases at high temperatures. Forging shapes the hot billet in the hot die. The lubricant (mold release agent, mainly graphite suspended in water or oil) prevents the billet from seizure and lengthens the die life.

2. The temperature control and the developed lubricant prevent the die from seizure.
3. The quantitative lubricant control maintains the quality of the forged pistons.
4. The billets (disk shape) for forging are prepared from a continuously cast bar which has a round cross-section.

The bar has a uniform microstructure without casting defects because of the well-managed casting process. This technology has made possible thin-walled pistons with accurate valve recesses. Figures 3.1 and 3.34 (left) are examples of controlled forged pistons.

3.6.3 Powder-metallurgical aluminum alloy raises high-temperature strength

The alloy composition suitable for conventional casting has been long established. AC8A and AC9B, typical piston alloys, have better heat resistance than common aluminum alloys. The high addition of transition elements such as Fe, Ni or Cr increases the quantity of intermetallic compounds, giving higher strength at high temperatures. However, too high Fe, Ni or Cr concentrations can form abnormally coarse intermetallic compounds, which cause brittleness. The high concentration also raises the melting temperature, making casting difficult. Hence, in conventional casting, it is difficult to improve heat resistance further. There is, however, a method to improve the heat resistance of aluminum alloy. This is to use powder metallurgy (PM).[24]

PM aluminum alloy is shaped by forging. The production process producing forging billets from an extruded rod is almost the same as that referred to in Fig. 2.24 of Chapter 2. In Fig. 2.24, a tube is extruded for the liner.[25] The method can be described as follows.

1. The molten alloy is sprayed into a powder. During solidification, the cooling rate measures above 10^3 °C/s. This rapid solidification makes the microstructure of the powder fine.
2. The conventional permanent mold casting generates coarse precipitation of intermetallic compounds which decreases strength. However, the rapid solidification increases strength with fine dispersions of intermetallics such as iron compound.
3. A hard particle such as SiC can mix with the alloy powder prior to the extrusion so that the wear resistance of the composite is adjustable.

The last row in Table 3.1 shows the chemical composition of such a processed PM piston alloy AFP1. High Si concentration is also chosen as a base metal. Fe and SiC are also added. Figure 3.36 compares the fatigue strength of the composite with that of AC9B.[26] AFP1 has a fatigue strength of 200 MPa at 10^7 load cycle at room temperature (20 °C). This means the alloy does not break below 200 MPa at 10^7 load cycle. It is about 50% higher than the strength of AC9B (135 MPa). The high strength results from the fine dispersion in the Fe-Al intermetallic compound at submicron level (Appendix G). In comparison with cast AC9B (Fig. 3.14), the Si in AFP1 distributes finely in a similar manner as that in Fig. 2.23 of Chapter 2. The fine distribution accommodates the microscopic stress concentration even at a higher stress level, restricting the initiation of fatigue crack. The fatigue strengths are 50% higher even at 150 and 250 °C (Fig. 3.36).

This improved strength enables much lighter weight designs. The PM alloy has excellent characteristics, but it needs to be forged. It is pointless to shape the piston by casting because re-melting the PM alloy eliminates the fine microstructure formed through rapid solidification.

Severe operational conditions can generate unfavorable deformation of the piston. Figure 3.37 shows dent deformation at the piston head during continuous operation at peak engine performance. The dent of the PM piston is less than that of the conventionally cast piston, showing excellent durability in a racing engine.[27] This is due to the high creep resistance of the PM alloy around 350 °C. These pistons have also been installed in mass-production snow mobile engines.[23] The piston weight has been lowered by 20% while the engine output has risen by 10%.

3.6.4   The iron piston

Iron base alloy is not suitable for use in a petrol engine piston. However, although very limited in number, iron pistons are used for diesels. At high cylinder pressure above 18 MPa, there is the potential to use cast iron or steel. The thin-walled design decreases weight and, for oil cooling of the piston head, a big cooling gallery compensates for low thermal conductivity.

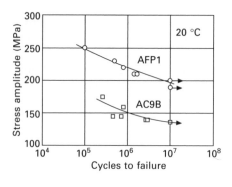

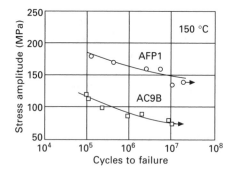

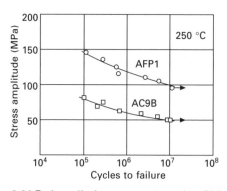

3.36 Fatigue limit curves comparing PM alloy and cast AC9B alloy. The vertical axis indicates the applied stress and the horizontal axis the cycles to fatigue failure. Failure does not occur at stresses below the curve.

Iron pistons can also reduce piston slap noise. When combined with a cast iron block, the small thermal expansion coefficient of iron alloys can decrease the clearance between the piston and the cylinder. The strength of cast iron at 350 °C is 16 times as high as aluminum alloy.

Typically, three types of iron pistons have been proposed for the market. These are, (i) a cast iron monolithic piston, (ii) a monolithic steel piston, and

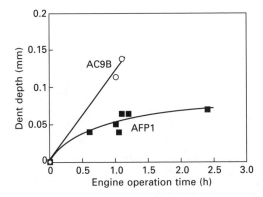

3.37 Piston head dents with engine operation time. The PM AFP1 piston is compared to the conventional cast AC9B piston. The dent of AC9B takes place faster than that of AFP1. Dent reduces the compression ratio. If it occurs during racing, the engine rapidly loses power.

(iii) a two-piece piston comprising a cast iron top and an aluminum alloy skirt. The third is called an articulated piston. This piston consists of two pieces, the skirt and piston head with a pin boss, which are connected to one another by the piston pin.

The cast iron piston uses spheroidal graphite cast iron having high strength. The steel piston uses medium carbon Cr-Mo steel. In the conventional design, the cast iron piston is 30% heavier than the aluminum piston but a shortened compression height can give nearly the same weight as the aluminum piston.[28] Quench-hardening heat treatment is used. All of these pistons have a big cooling gallery to cool the piston head. Figure 3.38 shows a steel piston that bonds the piston head with the skirt. To make the big cooling gallery, two pieces (head and skirt) are assembled by friction welding. Friction welding is described in Chapter 6.

3.7   Conclusions

The material used for a piston needs to fulfill a multitude of functions. A subtle balance between material property and heat management is essential for the design of a piston which generates high performance characteristics. The piston temperature of diesel engines is usually higher than that of spark ignition petrol engines. Since petrol has a lower flash point than the light oil used in diesels, too high a piston temperature can cause spontaneous ignition. If this occurs, the ignition plug cannot control combustion. In addition, the piston should be designed as light as possible because the petrol engine generates high power at high revolution. Therefore, light aluminum with high heat conductivity is preferable as a piston material.

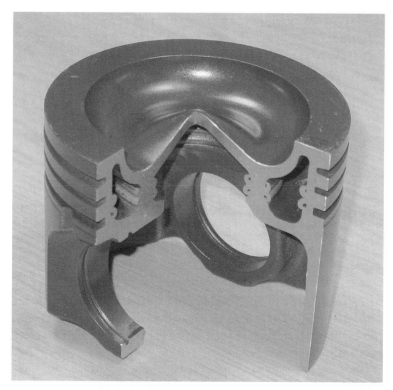

3.38 Friction-welded steel piston having a big cooling gallery (courtesy of Federal Mogul). The cut-away view shows the cross-section.

By contrast, diesel engines use compression ignition and can generate high power output at high cylinder pressure and low revolution. Although the exhaust gas temperature is lower than that of petrol engines, the higher the combustion temperature, the higher the efficiency. Iron base alloys are used for these reasons. However, if temperatures as high as 300 °C are continuously present at the cylinder bore surface or the piston ring area, then hydrodynamic lubrication by oil will not work. Hence, adequate cooling and low temperature at required portions are very important even in diesel pistons.

3.8   References and notes

1. Shioda W., *Keikinzoku*, 21 (1971) 670 (in Japanese).
2. Generally, two-stroke engine pistons have high operational temperatures, which are due to the difficulty of cooling. To raise seizure resistance more, there is AC 9A alloy containing 23% Si. This alloy has superior properties but is rarely used because of the difficulty in casting.

3. The coated piston with a solid lubricant is sometimes used in order to improve running-in wear see Table H.1 on page 293.

4. For example, die-casting can shape a thin-walled piston. However, it is difficult to use the piston at high temperatures. The high gas quantity generates blister defects during heating above 300 °C.

5. Ochiai Y., *Sousetsu Kikaizairyo*, Ver. 3, Tokyo, Rikougakusha Publishing, (1994) (in Japanese).

6. *Aluminiumno Soshikito Seishitsu*, ed. by Keikinzoku Gakkai, (1991) 248 (in Japanese).

7. There are two types of diamond bits: single-crystal bit and polycrystal bit. The polycrystal diamond bit is stronger in chipping than the single-crystal bit. However, the single-crystal bit gives a smoother cut surface. The polycrystal bit consists of many fine single crystals. Sintering using Co alloy bound the crystals.

8. *Jidousha Enjin Pisuton*, ed. by Suzuki Y., Tokyo, Sankaidou Publishing, (1997) (in Japanese).

9. Atake N., *Jidousha Gijutsu*, 47 (1993) 40 (in Japanese).

10. Koya E., *et al.*, Honda R&D Technical Review, 5 (1993) 43 (in Japanese).

11. Suto S., *et al.*, *Kinzoku Soshikigaku*, Tokyo, Maruzen Co., Ltd. (1972) 154 (in Japanese).

12. Suto H., *Kikai Zairyougaku*, Tokyo, Corona Publishing, (1985) 57 (in Japanese).

13. Yamagata H., *Yamaha Gijutsukai Gihou*, (1994) 3. The method to estimate a piston temperature above 350 °C is explained (in Japanese).

14. The difference between the estimated temperature and the measured temperature with a thermocouple is as small as 10 °C.

15. Furuhama S., *Nainenkikan*, 22 (1983) 61 (in Japanese).

16. Tomanik E., Zabeu C.B. and de Almeida G.M., SAE Paper 2003-01-1102.

17. Kurita H., *et al.*, SAE Paper 2001-01-0821. Anodizing for a high-strength piston alloy containing Cu was developed.

18. Donomoto T., *et al.*, SAE Paper 911284.

19. Yamauchi T. *et al.*, SAE Paper 830252.

20. This temperature, above which the strength rapidly decreases, is proportional to the melting temperature of an alloy. Generally, it corresponds to a half of the melting temperature.

21. Kolbenschmidt-Pierburg, Homepage, http://www.kolbenschmidt.de, (2003).

22. Koike T., *et al.*, *Proceedings of the 4th Japan International SAMPE Symposium*, (1995), 501.

23. Yamagata H. and Koike T., *Sokeizai*, 11 (1997) 7 (in Japanese). Yamagata H., *Proceedings of Towards Innovation of Superplasticity* II, JIMIS 9 Kobe, eds Sakuma T., *et al.*, Trans Tech Publications, (1999) 797.

24. Aluminum-base PM alloys are also used for compressors and cylinder liners. These parts require high wear resistance. Donomoto T., *et al.*, SAE Paper, 830252. Hayashi H., *et al.*, SAE Paper, 940847. The piston material requires wear resistance as well as strength in a wide temperature range.

25. PM aluminum alloy was first developed in Switzerland in 1945. Irmann R., *Metallurgia*, September (1952) 125. SAP (Sintered aluminum powder) alloy is well known. The extrusion technology developed in the 1980s has industrially enabled high quality materials.

26. The testing was carried out using over-aged specimens. The matrix strength decreases through over-ageing. However, the strength is kept high even at high temperatures

because the Fe-Al intermetallic compound in the AFP1 matrix does not coarsen even at high temperatures.

27. Koike T. and Yamagata H., *Proceedings of PM94 world congress*, Société Française de Metallurgie et de Materiaux and European Powder Metallurgy Association, Les Edition de Physique, (1994) 1627.

28. Federal Mogul catalogue, (2003).

4

The piston ring

4.1　Functions

The piston ring is essentially a seal with a spring-like property. Similar rings are also used in other piston and cylinder mechanisms, such as compressors or hydraulic devices. The piston ring of an internal combustion engine must be designed with sufficient heat resistance to withstand exposure to high-temperature gas. The single-piece metallic piston ring with self tension, which is generally used in internal combustion engines, was first invented by J. Ramsbottom in 1854.[1]

Figure 4.1(a) shows typical rings used in a four-stroke engine. The positions of the piston ring grooves have already been illustrated in Fig. 3.4 of Chapter 3. A piston generally uses three or more rings. Figure 4.2 shows a schematic representation of a piston, illustrating the relative positioning of the piston rings and the cylinder wall.

Figure 4.3 gives a summary of the various functions required of a piston ring. A running clearance of about 20–30 μm exists between the cylinder and piston. The piston rings seal in the combustion gas.[2] The combustion gas exerts pressure on the rings through the gap between the piston and cylinder bore (Fig. 4.2). The rings also control the thickness of the oil film on the cylinder wall, providing hydrodynamic lubrication which sustains a high thrust load. In addition, the rings also play an important role in cooling the piston head. The combustion heat received by the piston head flows into the cylinder wall through the piston rings. About 70% of the heat received by the piston head is transmitted through the rings. A high rotational velocity is necessary to generate high power output, requiring light piston rings with low friction and high wear resistance. Cast iron rings were often used in the past. However, more demanding requirements have increased the use of steel piston rings coated with various surface treatments.

Generally, four-stroke petrol engines use three rings per cylinder, whilst two-stroke engines use only two. Figure 4.1(a) shows the three rings of a four-stroke petrol engine, these being the top or compression ring (left),

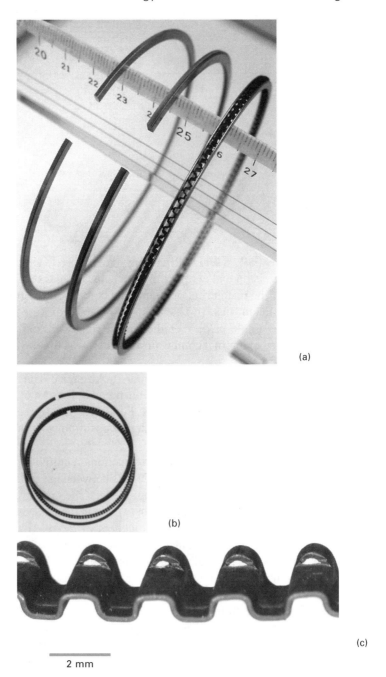

(a)

(b)

(c)

2 mm

4.1 (a) Piston rings for a four-stroke engine. Top and second rings (two rings on the left) and assembled three-piece oil control ring (on the right). (b) Disassembled three-piece oil ring. (c) Magnified view of the spacer. There is also a one-piece oil ring.

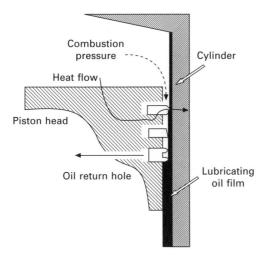

4.2 Phenomena taking place around piston rings.

second ring (middle) and oil control ring (right). The oil control ring consists of three individual pieces, two side rails and a spacer (the corrugated sheet, Fig. 4.1(c)). Figure 4.4 shows the two rings in a two-stroke petrol engine. The second ring is shown with the expander (located inside). The expander supports the second ring (described later in Fig. 4.9), adding tension without a significant increase in total weight. To obtain the same tension with a one-piece ring, the thickness needs to be increased, which in turn makes the ring much heavier.

Some diesel engines use more than three rings. In order to obtain high revolutions and quick response by reducing the weight of moving parts, fewer rings are preferred.[3] However, for more powerful engines with high cylinder pressures, such as diesels, a greater number of rings is required to obtain sufficient durability in sealing.

4.2   Suitable shapes to obtain high power output

Figure 4.5 illustrates a piston ring both before and after it expands into the ring groove. Figure 4.6 shows a ring installed in the ring groove. The piston with rings is inserted into the cylinder bore. The ring then expands from its initial diameter (d_1) and is forced tightly against the cylinder bore wall (Fig. 4.5). The ring width is called h_1 and the radial wall thickness a_1 (Fig. 4.6). The distance m is defined as the gap when the ring is uncompressed. The gap s_1, also referred to as the closed gap or end clearance, is the minimum gap obtained when the ring is installed in the cylinder bore. The load necessary to close the gap from m to s_1 is called the tangential closing force (Ft). The force increases by increasing the gap distance m. In the top ring of Fig. 4.1

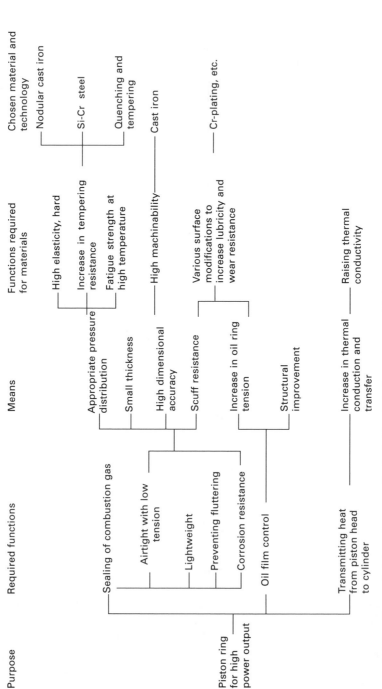

4.3 Functions of piston rings, particularly illustrated to generate high power output.

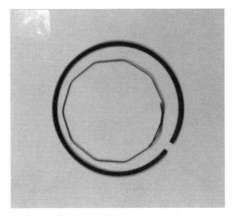

4.4 Piston rings for a two-stroke engine. The expander put at the center takes free state. When set into the piston ring groove, it spreads and gives additional force from the back of the second ring as shown in Fig. 4.9.

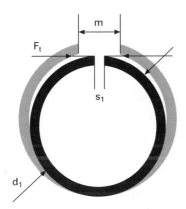

4.5 Nomenclature of a piston ring at open and closed states. The gap contracts from m (free gap size) to s_1 (closed gap, end clearance) when installed in the cylinder bore. The spacing between two facing planes forms a gap. This small portion including the gap is called 'butt ends'.

these values are typically $d_1 = 80$, $m = 10$, $a_1 = 3$ and $h_1 = 0.8$ mm, the ring being very thin to minimize weight.

It is the self-tension of the ring itself that presses the ring into the cylinder bore wall. During operation, the ring glides up and down, touching the bore wall. This puts stress on the ring. If the cylinder bore is not completely round and straight, the ring gap repeatedly opens and closes. The resulting stresses are likely to break the ring. A lack of lubrication also causes material failure. The surface roughness of the ring groove and degree of groove and side

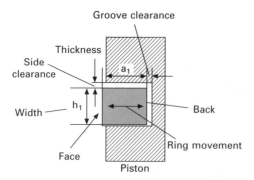

4.6 Cross cut view of a piston ring installed in the groove. The ring contacts the bore wall at the ring face. The inside surface against the ring surface is called ring back. The thickness is called a_1 and the width h_1.

clearances, are very important in controlling lubrication. Figures 4.7 and 4.8 show cross-sectional diagrams of three rings in a four-stroke engine and two rings in a two-stroke engine, respectively.[3]

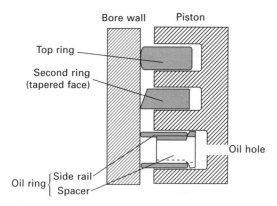

4.7 Three rings installed in piston-ring grooves for a four-stroke engine. The top ring has a barrel face shape. The oil control ring includes a sandwiched spacer between two side-rail sheets.

In four-stroke engines, the top (compression) ring is used mainly for sealing combustion gas. The second ring assists the top ring. The oil control ring is specifically used in four-stroke engines to scrape off lubrication oil from the bore wall. The second ring with a tapered cross-section also scrapes off the oil. The tapered face provides contact at the bottom edge to scrape oil during the downward stroke.

In two-stroke engines,[4] two rings are generally used without an oil control ring. (Fig. 4.8). The expander frequently supports the second ring (Fig. 4.9).

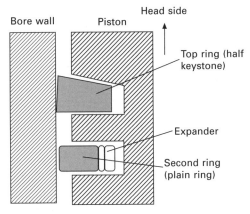

4.8 Two rings for a two-stroke engine. The top ring has a half keystone shape.

The tension created by the rings restricts the swing motion of the piston to suppress any abnormal stroke sound. Since increasing the a_1 size of a one-piece ring can make it much heavier, this two-piece construction raises the tension with less increase in total weight.

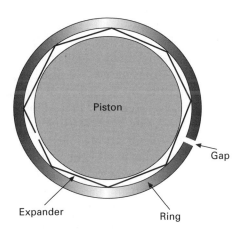

4.9 Expander installed at the back of the second ring of a two-stroke engine. Cross cut view at the second ring groove.

The ring motion follows the uneven shape of the cylinder bore wall. Both the distorted cylinder and the swing motion of the piston make the ring gap open and close repeatedly. During this motion, the degree of side clearance does not change for the rectangular type of ring (Fig. 4.10(a)), but it does for the keystone (wedge form) type of ring. Figure 4.10(b) illustrates the motion of a keystone ring. The keystone ring has the added benefit that it can

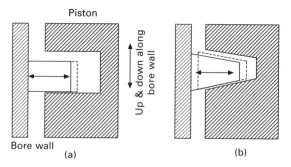

4.10 Section shapes of rings, (a) rectangle and (b) keystone.

eliminate accumulated dust such as soot in the ring groove. This cleaning prevents gumming up or sticking of the ring in the groove, which in turn decreases ring groove wear. Diesel and two-stroke petrol engines frequently use this type of ring. Half keystone rings (the top ring in Fig. 4.8) are also used in two-stroke engines. The keystone form is, however, more costly to produce.

A top ring with a barrel-shaped face (the top ring in Fig. 4.7) is frequently used. In maximizing lubrication, the shape prevents abnormal wear during the running-in stage and decreases blow-by. Ring fluttering can sometimes take place during increased revolution speeds and this increases blow-by. This is due to 'floating' of the ring. Floating occurs when an inertial force lifts the ring in the piston ring groove, which in turn spoils the airtight seal between the lower face of the ring and the ring groove. This can be dealt with by decreasing the ring weight by minimizing h_1. It is not feasible to decrease a_1 because it decreases contact pressure at the gap. Prevention of radial vibration can be achieved by either increasing a_1 or by using the pear type design shown in Fig. 4.17 which increases contact pressure.

Figure 4.11 illustrates typical designs of ring gap. Figure 4.11(a) is straight gap, which is the most standard shape in four-stroke engines. The sealing of the gap is very important. However, a minimum gap of about 0.3 mm is required to accommodate thermal expansion. While the engine is operating, this gap produces a very slight gas pressure leakage that could lead to ring flutter. Balancing the s_1 values of the top and second rings (gap balancing) can achieve a balance of pressures, so that the pressure between the top and second rings is never sufficient to lift the top ring from its seat on the bottom flank of the piston groove at the highest cylinder pressure. This gap balancing is required to minimize top ring flutter and its negative effects on cylinder gas sealing. Figure 4.11(b) shows a side notch gap with a locking pin hooking the semicircle edges together. This is used generally in two-stroke engines. There are other types such as a stepped gap design. These are effective, but very rare because the intricate machining is costly.

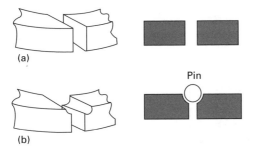

4.11 Gap shapes, (a) straight gap and (b) side notch gap. The piston ring should not rotate in the two-stroke petrol engine because the ports of the cylinder bore wall catch the gap (butt ends). Hence, a thin steel pin (locking pin) struck in the piston-ring groove, hooks the gap to stop the rotation.

4.3   Ring materials

4.3.1   Flaky graphite cast iron

Table 4.1 lists the various materials used in pistons. Two-stroke air-cooled engines use nodular graphite cast iron (JIS-FCD) for both top and second rings. Water-cooled engines use Si-Cr spring steel (JIS-SWOSC) for the top ring. Four-stroke engines use FCD or flaky graphite cast iron (JIS-FC) for second rings. The top ring and the side rail part of the three-piece oil control ring use SWOSC. The spacer of the oil control ring, the undulate sheet sandwiched between the side rail parts (Fig. 4.1(c)), requires a far more intricate shape, so it uses stainless steel JIS-SUS304 because of its good formability. The percentage of steel rings is increasing year by year. However, up until 1970, most engines used cast iron rings.

Piston rings are directly exposed to the very high temperatures of combustion gas, but they also receive heat from the piston head. The highest temperature appears in the top ring where temperatures reach about 250 °C. The material must maintain its elastic property at high temperatures for a long period of time.[5] Cast iron is excellent in this regard (Appendix D). A pearlite or tempered martensite microstructure (Appendices C and F) is generally used. Figure D.2 shows typical flaky graphite cast iron. The carbon crystallizes to generate flaky graphite during solidification of cast iron.

Cast iron has the following qualities that make it highly suitable for piston rings.

1. *Heat resistance.*   Cast iron rings are heat-resistant even when exposed to high temperatures. The hard martensite or pearlite microstructure does not soften at high temperatures. The high quantity of alloying elements (especially a Si content of around 3%) gives excellent resistance against tempering. Only casting can shape such high alloy compositions.

Table 4.1 Compositions (%) and applications of ring materials

Ring material	JIS	C	Si	Mn	P	S	Cr	Applications
Flaky graphite cast iron	FC	4	3	0.6	<0.2	<0.02	<0.4	4- and 2-stroke second rings
Nodular cast iron	FCD	4	3	0.6	<0.2	<0.2	–	4- stroke second ring. 2-stroke top and second rings.
Spring steel	SWOSC	0.5	1.4	0.7	<0.03	<0.03	0.7	4- and 2-stroke top and oil rings
Stainless steel	SUS304	<0.08	<1.0	<2.0	<0.04	<0.03	18(8Ni)	Oil ring spacer

Plastic working cannot shape cast iron into rings due to its low deformability.

2. *Self-lubrication.*   Graphite is self-lubricating, which helps to prevent scuffing. This is due to the layered crystal structure of graphite as described in Chapter 2. Scuffing[6] is a moderate form of adhesive wear characterized by macroscopic scratches or surface deformation aligned with the direction of motion. This is caused when the points on two sliding faces weld themselves together. Scuffing can occur between the cylinder bore wall and the ring or the piston outer surface.

3. *Machinability.*   Cast iron has good machinability. The dispersed graphite itself is soft and brittle, which works as a chip breaker during machining. A proper oil film must be produced between the ring face and cylinder bore wall. A residual burr at the ring corner is unfavorable, because it disrupts the oil film and obstructs hydrodynamic lubrication, thus all corners should be chamfered. Cast iron has high machinability compared to steel, which makes deburring much easier.

Sand casting is used to shape the flaky graphite cast iron ring. The distribution and shape of flaky graphite is very sensitive to solidification rate. Typically, a number of rings are cast together like a Christmas tree as illustrated in Fig. 4.12. This casting plan hangs several rings around the downsprue and runner, and ensures that all the rings of one tree will have a homogeneous graphite distribution.

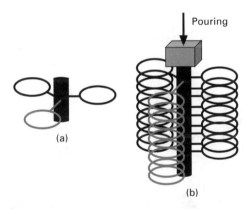

4.12 Casting plan for flaky graphite cast iron rings, (a) rings produced by one layer of the mold and (b) rings produced by the stacked mold.

An alternative method is to slice a cast iron tube into rings. It may be cheaper, but this method gives various solidification rates at different portions of the tube, which in turn disperses graphite unevenly. Hence, particularly for flaky graphite cast iron, each ring should be cast separately. High-alloy cast iron is used to give much higher wear resistance. It disperses Cr-carbide through increasing Cr content or hard steadite (iron-phosphorus compound, see Chapter 2) through an increased phosphorus quantity of around 0.3%.

4.3.2   Use of spherical graphite cast iron to improve elastic modulus and toughness

Decreasing h_1 can make the cast iron ring lighter, but it also raises the stress. Cast iron has excellent properties as a ring material, but is not that tough. The microscopic stress concentration caused by flaky graphite is likely to initiate cracking, and the flaky graphite microstructure is too weak to resist such cracking. To increase the strength, nodular graphite cast iron (JIS-FCD), which includes spherical graphite, has become more widely used. It is also called spheroidized graphite iron or ductile iron, as mentioned in Chapter 2. This microstructure is resistant to cracking. Figure 4.13 is a magnified view showing tempered martensite surrounding spherical graphite in the second ring of a two-stroke engine. Figure 4.14 is a photograph of the ring in cross-section. This is a half keystone shape with hard chromium plating on its face. The hardness is around 40 HRC due to the increased concentrations of Cu, Cr and Mo.

The tempered martensite microstructure of flaky graphite iron gives a bending strength of 400 MPa and an elastic modulus of 100 GPa, while that

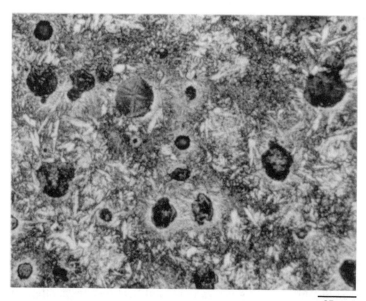

25 μm

4.13 Nodular graphite cast iron with martensite matrix.

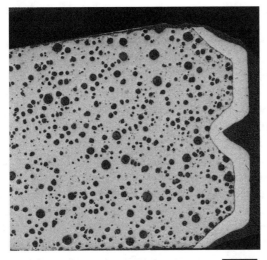

200 μm

4.14 The ring section of a two-stroke second ring. The magnified view is shown in Figure 4.13 The groove at the middle of the plated chromium layer is a scuff band. Even if slight scuffing takes place, it prevents the wear scar from extending to the entire face and keeps sealing.

of nodular graphite iron gives 1.2 GPa and 166 GPa, respectively. Hence, spheroidizing substantially improves mechanical properties. Round graphite is generated by adding small amounts of nodularizer to the melt just before pouring.[7] The nodularizer is a Mg and/or rare earth Ce alloy containing Si, Fe and Ni. This processing originated with simultaneous inventions in 1948. J.H. Morrogh discovered the spheroidizing effect of adding Ce, and A.P. Gagnebin through adding Mg. Nodularizer is widely used to increase the strength of cast iron through adjusting the geometrical shape. For a nodular graphite iron ring, manufacturing starts from a cast tube. The ring is then sliced from the tube and the gap is notched. The machined ring is quench-tempered to create the necessary tension.

More recently, the use of steel rings has been increasing. However, the second ring of four-stroke cycle engines generally uses JIS-FC or FCD cast iron without chromium plating, because it is difficult to grind steel into the required tapered face (Fig. 4.7).

4.3.3   Using steel to generate lightweight rings

Up until 1970, all piston rings were made using cast iron. However, the low fatigue strength and toughness of cast iron mean that it is not possible to reduce the weight of the rings by lowering h_1. Steel rings[8] using spring steel have been developed to address this problem. Steel does not have the self-lubricating property of cast iron, but it does have excellent elastic properties. Various spring steels have been tried for piston rings. At present, Si-Cr steel, which is also used for valve springs, is widely used because of its high resistance to tempering. The typical chemical composition is Fe-0.5% C-1.4Si-0.7Mn-0.7Cr (Table 4.1). It is generally used with a tempered martensite microstructure. The high Si content maintains the hardness of the martensite in the middle to high temperature range, which in turn maintains ring tension.

Presently, nearly half of all piston rings manufactured use steel, and this is likely to increase further still in the near future. The use of a steel second ring is also becoming more common, despite difficulties in machining the taper face. Figure 4.15 illustrates the manufacturing process of a steel ring. First, rolling produces a wire with a rectangular section (upper left). This wire is then coiled into an oval shape (1) so that the final shape after installation is round. Quench-tempering (3) generates the required elastic property (described below). The tensile strength after quench-tempering is typically 1.5 GPa, and the elastic modulus 206 GPa. After heat treatment, a lapping machine (illustrated in Fig. 4.16) generates a barrel shape (4) from the

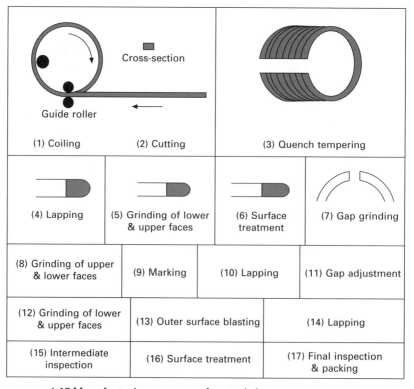

4.15 Manufacturing process of a steel ring.

rectangular cross-section. A cylindrical whetstone laps the outer surface of the stacked rings. The shaft revolves and moves up and down with the rings. During this motion, the rings are slightly inclined in the cylindrical whetstone to generate the barrel face.

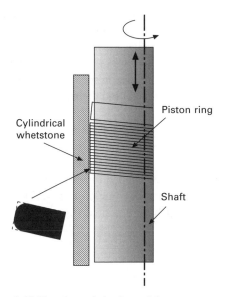

4.16 Shaping of the barrel face.

Each corner requires a different radius for lubrication. Simple deburring, such as barrel polishing, cannot be used because the accuracy of the chamfer after barrel polishing cannot be controlled. Grinding and lapping should be carried out on every corner and face. Deburring steel is much more difficult than deburring cast iron, because of its high ductility.

The top ring for diesel engines is exposed to a much higher temperature and pressure than that of petrol engines. In addition to the Si-Cr steel, the diesel engine frequently uses high-chromium martensitic stainless steel (17% Cr steel containing Mo, V, etc.) with additional nitriding (described below in Table 4.2), which shows superior anti-softening properties at high temperatures.

The steel ring with its high elastic modulus is also beneficial in terms of weight reduction, but it is not always the best solution. For example, compared to cast iron, the a_1 of the steel ring should be lowered to adjust the ring tension, but then the contact area between the ring and ring groove decreases, reducing heat transfer. A cast iron ring with a lower elastic modulus is much more favorable in such a case.

4.4   Designing the self-tension of rings

4.4.1   The distribution of contact pressure and tension

Higher contact pressure for the rings is essential at higher-speed revolutions. This is because the hydrodynamic force that occurs in the oil film and tends to lift the ring away from the cylinder wall increases with sliding velocity. The self-tension of the ring forces it against the bore wall, which in turn generates a contact pressure. The combustion gas pressure transmitted through the groove clearance also forces the rings towards the bore wall, helping to increase the contact pressure. However, at high piston speeds, the time required for the formation of an effective gas pressure behind the rings becomes much shorter.

Figure 4.17 shows an example of the radial pressure pattern of piston rings. The black line has a peak contact pressure at the gap. This distribution is unacceptable, because such a localized high pressure is likely to disrupt the oil film. For four-stroke engines, a pear-shaped distribution with a fairly high value at the gap is ideal (shown by the gray line on Fig. 4.17). The most suitable shape for contact pressure distribution is determined by the engine type and material properties.

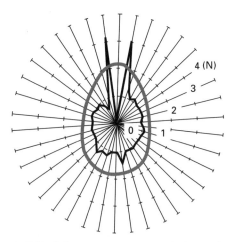

4.17 Radial pressure pattern of piston rings measured by a pressure sensor, contact pressures indicated by load (N). The gap locates at the top position. The gray line illustrates a favorable shape.

Designing a piston ring begins with calculating the contact pressure distribution. The following factors should also be taken into consideration: preventing blow-by, minimizing oil consumption, and decreasing friction loss and wear. These factors all determine the dimensions of the ring.

In a rectangular section ring, the mean specific contact pressure P (MPa) is calculated; $P = E \ (m_1 - s_1)/d_1/(7.07(d_1/a_1 - 1)^3)$, where E (MPa) is the Young's Modulus and the dimension of every part is given in mm. The tangential closing force Ft (N) that acts upon the ends, i.e., the force which is necessary to press the ring together at the end clearance, is calculated; $Ft = P \cdot h_1 \cdot d_1/2$. The stress f (MPa) that the ring material receives is; $f = E \cdot a_1 \ (m - s_1)/2.35/(d_1 - a_1).^2$

The top ring and second ring of four-stroke petrol engines have a contact pressure of around 0.2 MPa, the oil ring in the range from 0.8 to 1 MPa. The oil ring for diesel engines has a contact pressure ranging from 1.6 to 2 MPa. The average tension Ft changes with the dimensions of the ring. The combustion gas forces the top and second rings towards the bore wall, but does not push the oil ring against the bore wall because combustion gas leakage is sealed almost perfectly by the top and second rings. Hence, tension in the oil ring is generally high.

Rings with higher contact pressure remain more effective over longer running periods. The stress loss from wear reduces contact pressure, whilst a high initial value of contact pressure tends to result in greater residual stress than the low initial value found in low-tension rings.

Lack of oil causes severe wear of the ring and bore wall, while excess oil generates too much soot. Soot accumulates in the combustion chamber and causes combustion conditions to deteriorate, which can result in a number of problems, including a tendency for the ring to stick to the ring groove. This is partially eliminated by using a keystone ring, but optimum oil control is still necessary. The quantity of oil is adjusted mainly by the oil control ring, although the combined effects of all rings, including the compression ring, should be taken into consideration.

Operating conditions also influence oil consumption. The number of revolutions has a significant influence, as does the negative pressure inside the inlet pipe during engine braking. Increased tension in the oil ring rapidly leads to lower oil consumption, but flexibility is also important. Constant oil consumption appears above a certain tension value, and the correct tension value is determined empirically.

More recently, reducing fuel consumption has become important. To accomplish this, friction must be reduced and moving parts must be lighter. It has been said that the friction loss arising from the piston, piston rings and cylinder bore amounts to 40–50%[9] of total friction loss in engines. For piston rings, friction on the cylinder bore is reduced by lowering the contact pressure and by using a narrower ring width. But if the contact pressure is reduced, sealability and oil consumption cannot be maintained, and the roundness and straightness of the bore must also be taken into account.

4.4.2   Tensioning

The most common designs of piston rings have a non-circular shape in the free state, so that when they are installed, they will conform tightly to the cylinder wall at every point and the desired contact pressure distribution will be obtained. This favorable shape, characteristic of the ring in its free state, can be produced by several different processes.

In the case of cast iron rings, the shape of the casting gives the desired contact pressure distribution. The flaky graphite iron is shaped into an oval without a gap (Fig. 4.12). In nodular iron rings, the tube cast for slicing is oval in cross-section, and the circular shape is obtained by cutting the gap. During processing, the gap is subjected to several repeated cycles of opening and closing. This load cycle on the ring removes micro-yielding (anelasticity) to increase elastic properties. Micro-yielding is the phenomenon where small plastic deformation takes place before the macroscopic elastic limit is reached (Appendix K). The repeated load cycle on the ring removes it. This effect is very important and is called accommodation.

In steel rings, the shape generated in the coiling process (Fig. 4.15) determines contact pressure distribution geometrically. It is also possible to grind a non-circular shape out of a circular ring, but this raises the cost quite considerably. An alternative method of generating an oval shape involves first coiling the wire onto a circular form. The coil is then pressed around a core bar with an oval section. Heat treatment causes the coil to deform thermally and conform to the oval shape of the bar. This process is called thermal tensioning and is also applied to cast iron rings.

4.5   Surface modification to improve friction and wear

4.5.1   Surface modifications during running-in

Ring wear usually appears at the outer, top and bottom faces. However, the rings are not all subjected to the same conditions. The conditions are most severe for the top ring since it is directly subjected to the high pressure, high temperature and considerable chemical corrosive effects of combustion gas. Furthermore, the top ring also receives the lowest supply of lubricating oil. On the cylinder bore, the upper reversal point of the top ring (top dead center) is likely to suffer the greatest amount of wear.

It is important to improve the tribological properties of rings.[10] Table 4.2 summarizes typical surface treatments used on ring materials (more detail is given in Appendix H). The surface treatments can be classified into (i) improving initial wear during running-in, and (ii) improving durability where very long running distances are required (for instance, for commercial vehicles, which are expected to run for several hundred thousand kilometers).

Table 4.2 Ring materials with surface treatments. Gas-nitrided martensitic stainless steel generates a hard nitrided surface layer containing carbides and nitrides, which has superior wear and scuffing resistance. Physical vapor deposition (PVD) is a coating process in which a vapored metal is deposited under a reduced pressure atmosphere. CrN is widespread for piston rings. Ionized Cr is deposited under nitrogen gas at high adhesion speed (Appendix H)

Specifications		Base material	Outer surface modification	Side surface modification
Top ring		Nodular cast iron	Cr plating	Phosphate conversion
		Si-Cr steel	Cr plating	Fe_3O_4 coating, Solid lubricant coating
		Martensitic stainless steel	Gas nitriding, Composite plating, Physical vapor deposition	Phosphate conversion Solid lubricant coating
Second ring		Nodular cast iron	Cr plating	Phosphate conversion
		Gray cast iron	Phosphate conversion	Phosphate conversion
Oil control ring	3-piece	Side rail	Carbon steel	Cr plating	Fe_3O_4 coating, Phosphate conversion
		Martensitic stainless steel	Gas nitriding Ion nitriding	Phosphate conversion
		Space expander	Austenitic stainless steel	Salt bath nitriding	
	2-piece	Oil ring surface piece	Carbon steel	Cr plating	Fe_3O_4 coating Phosphate conversion
		Martensitic stainless steel	Gas nitriding	
		Coil expander	Carbon steel	Cr plating	
		Austenitic stainless steel	Salt bath nitriding	

The running-in period is important, but unfortunately, drivers cannot always be relied on to conform to running-in requirements. To counteract this, surface treatments have been developed specifically for the running-in period. Typically, phosphate conversion coating, as listed in the table, is used. This is a chemical conversion treatment[11] that generates a phosphate film, for instance manganese phosphate, through dissolving the iron substrate. The coated layer is porous, soft and insoluble, and retains oil to improve the initial accommodation between the parts. The treatment also removes burrs by dissolving the substrate, and prevents rust.

4.5.2   Surface modifications to improve durability

Combustion products generated inside the engine cause abrasive wear, as can the dust contained in the intake air or wear debris from the various parts. Without a hard surface coating, steel rings have poor resistance to scuffing. To improve durability, various materials are used to coat not only steel rings but also cast iron rings. Hard chromium plating is widely used to increase the wear resistance of the ring face.[12] It has a hardness of about 800–1200 HV, while cast iron rings have a hardness of about 220–270 HV. The plating is about 50 μm thick and considerably increases wear resistance. Figure 4.18[13] compares the amount of wear with or without chrome plating as a function of friction velocity. The chromium-plated cast iron shows less wear.

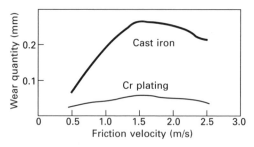

4.18 Friction velocity vs. degree of wear of cast iron specimens with or without chromium plating. The testing machine is Kaken type. The partner material is also a flake graphite cast iron. The contact load is 50 MPa and the friction distance is 10 km at room temperature under dry air.

The hard chromium plating used for piston rings is porous chromium plating, as discussed in Chapter 2. The crack density included in the plated layer is carefully controlled so that the cracks retain adequate lubricating oil. More recently, composite chromium plating, which includes 4–11% Al_2O_3 or Si_3N_4 powder, has been proposed.[14] It has a higher wear resistance than conventional hard chromium plating. First, the ceramic powder is mechanically forced into the crack in the plated layer, then the additional chromium is overlaid. This two-stage plating process makes a composite coating. A composite Ni-Si$_3$N$_4$ plating including phosphorus is also used to raise scuff resistance.

Chrome plating is far more effective and generally less costly, but the associated liquid waste contains the hexavalent Cr ion, which is a health and safety risk. An alternative dry treatment, such as CrN physical vapor deposition (PVD), which does not use an electrolyte, has been brought into wide use. CrN has a faster deposition rate than hard TiN, which is popular in cutting tools (Appendix H). The hardness is in the range of 1,600 to 2,200 HV. Figure 4.19 is a schematic representation of a PVD chamber. The vaporized metal positioned at the cathode is ionized and then deposited onto the stacked piston ring. The chamber is controlled under the reduced pressure of the process gas. The table is rotated to provide a uniform film thickness. In addition to this coating method, metal spray using Mo can also be used.

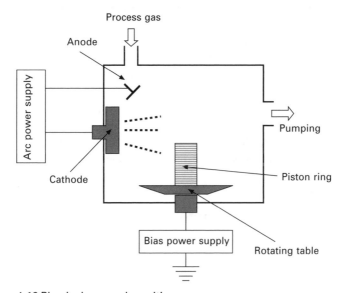

4.19 Physical vapor deposition.

The relative scuffing resistance of four other coatings is compared with that of hard Cr plating in Fig. 4.20. In this comparison, PVD-CrN coating is the most scuff resistant. It has been reported that CrN including oxygen shows the highest resistance,[15] however, relative performance depends on test conditions. Another coating aimed at reducing friction loss tried recently uses a carbon-based film, typically diamond-like carbon.[15] In tests using a reciprocative wear testing machine, the friction coefficients decreased in the following order: hard-chromium plating, nitrided stainless steel, PVD-CrN

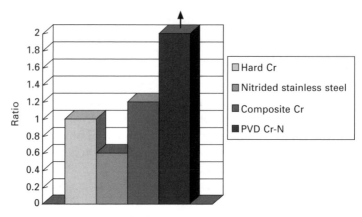

4.20 Comparison of scuffing resistance.

and diamond-like carbon. The coatings used are selected carefully for each engine, taking durability, cost and other factors into account.

4.6   Conclusions

High power output frequently causes problems around the piston and piston ring. For example, piston scuffing takes place at high engine temperatures and is caused by insufficient cooling. Soot resulting from inappropriate combustion causes scuffing in the ring grooves. The piston, piston ring and cylinder all act together to generate the required performance. Mechanical design that allows for appropriate lubrication must be considered before final material selection.[16]

4.7   References and notes

1. *Jidoushayou Pisutonringu*, ed. by Jidoushayou Pisutonringu Henshuiinkai, Tokyo, Sankaido Publishing, (1997) (in Japanese).
2. Some small miniature engines do not use piston rings at all. The small bore diameters of these engines can maintain a small running clearance between the piston and cylinder even at elevated temperatures. This minimizes blow-by without causing piston seizure.
3. There is an attempt to decrease the ring number, two rings in four-stroke engines and one ring in two-stroke engines. The purpose is to decrease the inertial weight. However, less durability is likely to increase blow-by and oil consumption. Presently, some high performance engines use this construction.
4. Two-stroke engines can supply less lubrication oil around the piston ring. The combustion frequency is twice that of four-stroke engines, so that the piston ring is continuously pressed to the ring groove bottom. By contrast, the lubrication oil in four-stroke engines thrusts into the clearance between the ring and ring groove. The ring lifts in the groove with the inertial force during the exhaust cycle.

5. JIS-B8032 prescribes that the decline of the ring tension, caused in the cylinder bore for one hour at 300 °C, should be below 7% in JIS-FC and 10% in JIS-FCD rings.

6. Ebihara K. and Uenishi J., *Pisutonringu*, Tokyo, Nikkankougyou Shinbun Publishing, (1955) 135 (in Japanese).

7. Cho H., *et al.*, *Kyujoukokuen Chutetsu*, Tokyo, Agne Publishing, (1983) (in Japanese).

8. The ring was functionally divided into the compression ring and oil control ring in 1915. The steel ring was first used for oil rings in 1930. Tomitsuka K., *Nainenkikanno Rekishi*, Tokyo, Sanei Publishing, (1987) (in Japanese).

9. Yoshida H., *Tribologist*, 44 (1999)157 (in Japanese).

10. Not only the piston ring but also the piston-ring groove should have enough wear resistance. Hard anodizing is usually applied for the top ring groove. See Chapter 3.

11. Bonderizing used for cold forging is also a similar chemical conversion. It is a Zn-phosphate coating permeated with soap. The Mn-phosphate coating is commercially called parkerizing. Unlike plating, the phosphate conversion dissolves iron substrate. See Appendix H.

12. Decoration-chrome plating aims at corrosion resistance and luster. We generally see decoration-chrome plated goods in the market. For decorative purposes, cracks in the plated layer are unfavorable. Water penetrates into the iron substrate through the cracks to cause rust. To prevent this, a thin copper or nickel layer is plated first, and then the chrome is overlaid.

13. Riken Co., Ltd, *Piston Ring and Seal Handbook*, (1995) (in Japanese).

14. Miyazaki S., *Tribologist*, 44 (1999)169 (in Japanese).

15. Iwashita T., *Tribologist*, 48 (2003)190 (in Japanese).

16. Furuhama S., *Jidousha Enjinno Toraiboloji*, Tokyo, Natsume Publishing, (1972) (in Japanese).

5.1 Functions

Combustion gases in four-stroke engines are controlled by the valve mechanism, a complex structure, often referred to as a valve train, of which the camshaft is an integral part. The valve train determines overall engine performance. Figure 5.1 shows a photographic representation of a valve train. Table 5.1 lists the main parts of a valve train and the typical materials used in

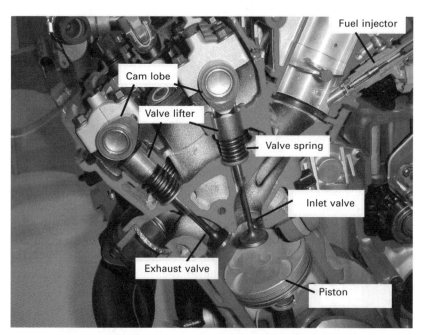

5.1 DOHC type valve system. The valve and cam lobe are not connected. The valve spring presses these two parts together. The valve lifter (bucket tappet) is positioned in between the valve stem end and cam lobe.

Table 5.1 Typical materials in valve trains

Part name	Material
Camshaft	Chilled cast iron, hardenable cast iron or JIS-SCM420 forging followed by carburizing
Valve lifter	JIS-SKD11 cold forging (quench-temper)
Rocker arm	JIS-SCM420 forging (carburizing) + Cr-plating or + wear-resistant sintered-material chip (brazing)
Inlet valve	Martensitic heat-resistant steel JIS-SUH3 forging
Exhaust valve	Austenitic heat-resistant steel JIS-SUH35 (crown) forging + JIS-SUH1 or SUH3 (shaft), friction welded. Stellite hardfacing on the valve face
Inlet valve sheet	Iron-base heat- & wear-resistant sintered-material (press-fit into the cylinder head)
Exhaust valve sheet	Iron-base heat- & wear-resistant sintered material (press-fit into the cylinder head)
Valve spring	Si-Cr steel oil-tempered wire + shot peening

its construction. Most of these parts are produced from high carbon iron alloys.

The valve train consists of a valve operating mechanism and a camshaft drive mechanism. The valve operating mechanism transforms rotation of the crankshaft into reciprocating motion in the valves. The valves protrude into the combustion chamber and are pushed back by the reactive force of the valve spring.

Several types of valve trains have been developed. The overhead camshaft is the most popular mechanism used in high-speed engines. There are two types, the double overhead camshaft (DOHC) and single overhead camshaft (SOHC). Figure 5.1 shows an example of the DOHC type, which uses five valves per cylinder (two exhaust and three inlet valves). This mechanism uses two camshafts, one camshaft drives the three inlet valves and the other drives two exhaust valves through the valve lifters.

Figure 5.2 gives a schematic representation of a typical DOHC drive mechanism. The chain or timing belt transmits the rotation of the crankshaft to the camshaft, which is turned by the camshaft drive mechanism. Figure 5.3 shows the SOHC type. This mechanism uses one camshaft, which drives a pair of inlet and exhaust valves via the rocker arms.

An example of a camshaft is shown in Fig. 5.4. The functions of the camshaft are analyzed in Fig. 5.5. The camshaft turns at half the rotational speed of the crankshaft, which is synchronized by the crankshaft rotation. If the number of revolutions is 12,000 rpm at the crankshaft, then the camshaft turns at 6,000 rpm, resulting in reciprocating motion of the valves at 6,000 times a minute.

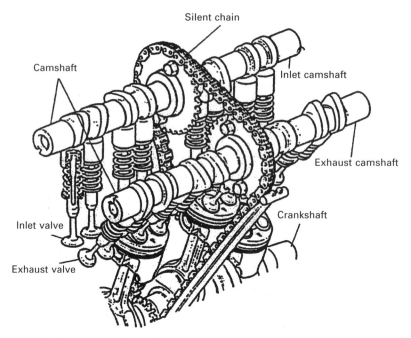

5.2 Schematic illustration of a valve train.

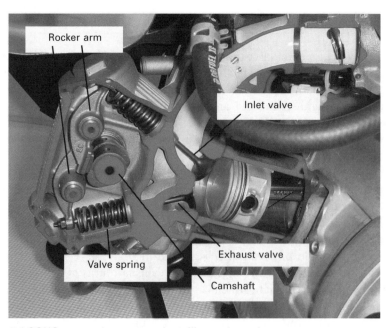

5.3 SOHC type valve system installing twin rocker arms.

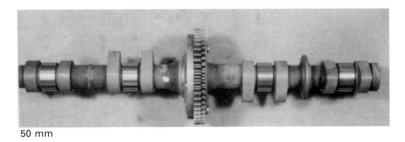

50 mm

5.4 Camshaft installing a drive sprocket at the center. The cam lobe converts rotation into reciprocating motion.

The oval shape of the cam lobe determines the lift (displacement) of inlet and exhaust valves. The valve itself has an inertial mass. If the curved shape of the cam lobe surface is not designed appropriately, then the valve cannot accurately follow the contour and this will result in irregular motion. This is likely to occur at high revolutions. Lighter moving parts in the valve train will enable high-speed revolutions. Increasing the tension of the valve spring will increase reactive force, helping to prevent irregular motion of the valves. However, the high reactive force will result in high contact pressure on the cam lobe, so the cam lobe should have high wear resistance.

It is essential that adequate amounts of lubricating oil are supplied to the cam lobe. The contact between the curved surface of the cam lobe and the flat face of the valve lifter (bucket tappet) generates high stress,[1] and therefore both parts require high wear resistance where contact occurs. In the DOHC mechanism, the cam lobe makes contact with the head of the valve lifter directly or via a thin round plate (pad or shim), which is positioned on the valve lifter head. The high contact pressure means a much harder material is needed for the shims.The SOHC mechanism uses rocker arms (Fig. 5.3). The face that is in contact with the cam lobe also needs to have good wear resistance.

5.2   Tribology of the camshaft and valve lifter

The reactive force of the valve spring must be set high in order to maintain smooth motion and generate high revolutions, as discussed above. The maximum permissible surface pressure, usually regarded as the decisive parameter limiting cam lobe radius and the rate of flank-opening, currently lies between 600 and 750 MPa, depending on the materials used.[2]

When the camshaft is operating at high revolutions, contact pressure is reduced by the inertia of the valve lifter. Under these conditions, the oil film on the running face is maintained most easily, providing hydrodynamic lubrication. Contact pressure is therefore highest and lubrication most challenging when the engine is idling. Figure 5.6[3] summarizes the basic relationships between the factors that influence the tribology of the camshaft

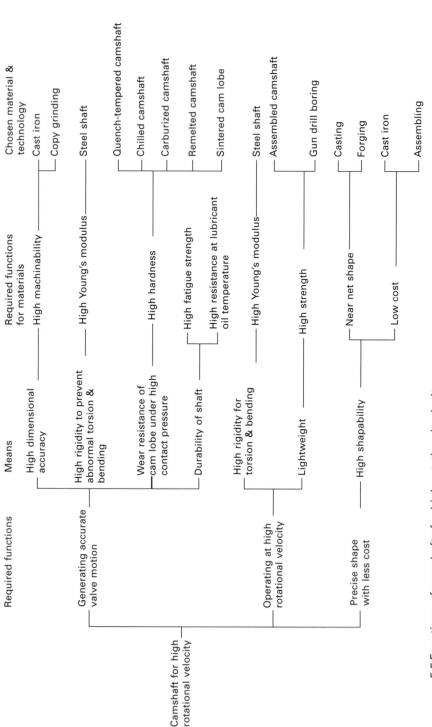

5.5 Functions of camshafts for high rotational velocity

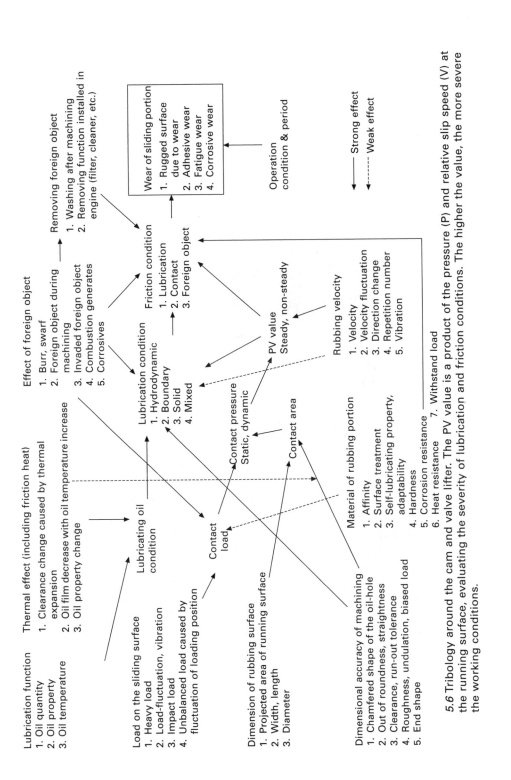

5.6 Tribology around the cam and valve lifter. The PV value is a product of the pressure (P) and relative slip speed (V) at the running surface, evaluating the severity of lubrication and friction conditions. The higher the value, the more severe the working conditions.

and valve lifter, and which can therefore cause problems that result in wear at the point of contact.

Figure 5.7 shows an example of flaking at the head of a DOHC valve lifter. Flaking is caused by surface fatigue. The Hertzian stress reaches its highest value just under the contact surface, frequently resulting in fatigue cracks that then cause flaking (see also Chapter 9). In Fig. 5.7, the surface has peeled off to reveal the cavities underneath, a typical failure under high contact pressure.

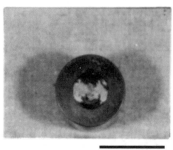

20 mm

5.7 Flaking appearing in the valve lifter head. Flaking is a type of wear where the face comes off like a flaky powder.

Pitting is another surface fatigue phenomenon. Pitting normally manifests itself as small holes and usually appears under high contact pressures. Figure 5.8[3] summarizes the main reasons why pitting occurs in the cam lobe and the factors that affect each of these reasons. The increased temperature at the running surface that results from increased friction lowers the viscosity of the lubricating oil, making it less efficient. Under these conditions, the mating metal surfaces lose their protective oil film and come into direct contact. Wear can appear on either the tappet or the cam lobe. It is very important to choose an appropriate combination of materials.

The function of the shaft itself is also very important. The torque from the crankshaft drives the camshaft, so the shaft portion is under high torque and therefore must have high torsional rigidity. Figure 5.9 shows a section taken at the journal-bearing portion (as indicated in Fig. 5.5). The hole at the center runs along the entire length of the camshaft and supplies lubricating oil to the journal bearings.

5.3   Improving wear resistance of the cam lobe

5.3.1   Chilled cast iron

The camshaft should combine a strong shaft with hard cam lobes. Table 5.2 lists five types of camshaft.[4] Table 5.3 lists the chemical compositions of the

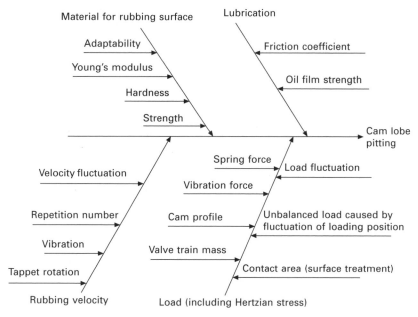

5.8 Reasons causing pitting. A tappet is the counterpart of the cam lobe in an overhead valve engine.

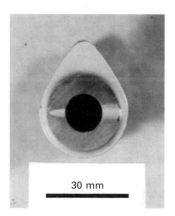

5.9 Camshaft cross-section at the position of an oil hole which is perpendicular to the central hole. The holes supply oil to the journal bearing.

various materials used. The most widely used material for camshafts at present is chilled cast iron ((1) in Table 5.2), using a high-Cr cast iron. This type of camshaft is shown in Fig. 5.4, and has hard cam lobes with a strong but soft shaft.

The chilled camshaft utilizes the unique solidification characteristics of cast iron. Figure 5.10 illustrates the production process. Let us consider a

Table 5.2 Various camshafts. The counterpart of the camshaft uses a forged steel plated by hard chromium, a sintered metal chip dispersing carbide or a nitrided JIS-SKD 11 plate. The dispersed carbide in the chilled cam lobe gives excellent wear resistance

Type	Cam lobe portion	Shaft portion	Processing	Characteristics
(1) Chilled cam	Chill	Flaky or spherical graphite cast iron	Sand casting combined with a chiller	Most general. Hardness control is difficult
(2) Remelted cam	Chill	Flaky or spherical graphite cast iron	Remelting the cam lobe surface of the shaped material of gray cast iron	Increasing the hardness of the cam edge portion is difficult
(3) Quench-tempered cam	Martensite	Quench-tempering or normalizing	Quench-hardening the cam lobe by induction or flame heating	Applicable to forged carbon steel, nodular cast iron or hardenable cast iron
(4) Carburized cam	Martensite	Sorbite	Carburizing the forged part (SCM 420)	Strong shaft portion using a thin wall tube
(5) Bonded cam	Wear-resistant sintered material Martensite	Steel tube	Brazing, diffusion bonding or mechanical joining of the cam lobe with the shaft	Flexible choice and combination of various materials

Table 5.3 Compositions of camshaft matrials (%). The high-Cr cast iron is used for chilled camshafts. The chromium concentration is slightly raised to obtain hard chill. The hardenable cast iron generates a martensitic microstructure through quench-tempering. The Cr-Mo steel SCM 420 is forged and carburized. The sintered metal has a martensitic microstructure dispersing Cr and Fe complex carbide

Material	C	Si	Mn	Cr	Mo	Cu	V	W	Fe
High-Cr cast iron	3.2	2.0	0.8	0.8	0.2	–	–	–	Balance
Hardenable cast iron	3.2	2.0	0.8	1.2	0.6	–	–	–	Balance
Cr-Mo steel JIS-SCM420	0.2	0.3	0.8	1.0	0.2	–	–	–	Balance
Sintered metal for cam lobe	0.9	0.2	0.4	4.5	5.0	3.0	2.0	6.0	Balance

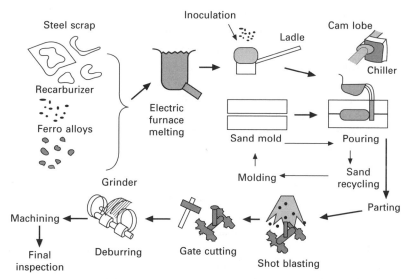

5.10 Casting process. First, the electric furnace melts steel scraps, carbon content raiser (carbon powder) and ferro-alloys (Fe-Si, Fe-Cr alloys, etc.). Then the melt taken in the ladle is poured into the mold. The mold is a sand mold. A chiller is inserted in advance at the cam lobe position where chilled microstructure is required. After solidification, the sand mold is broken and the camshaft is taken out. The sand mold contains a binder and appropriate water content. It should be breakable after solidification without break during pouring. The sand is reused. The iron shots blast the shaped material to remove the sand. The unnecessary gate, sprue and runner are cut. The remnants are remelted and reused. Grinding deburrs the shaped material. Then it is directed to the final machining. Casting is an extremely rational production process.

gradual increase in carbon concentration towards 4.3% in the iron-carbon phase diagram. (Please refer to Appendices C and D for more detail on the phase diagram of the iron-carbon system.) Pure iron solidifies at 1,536 °C. The solidification temperature decreases with increasing carbon concentration to give a minimum value of 1154 °C at a carbon concentration of 4.3%, the eutectic point, (see Fig. C.1).

Molten iron is transferred from furnace to molds using a ladle covered with a heat-insulating lining. In manual pouring, one ladle of molten iron can be poured into several molds one after another, which takes around five minutes. If the solidification temperature of the metal is high, then the pouring must be finished within a very short period of time otherwise the molten iron will solidify in the ladle. Hence, with a lower solidification temperature there is more time for pouring.

Sand molds produce a slow solidification rate because the insulating effect of the sand slows cooling. Under these conditions, the carbon in the cast iron crystallizes as flaky graphite (Fig. 5.11(a)) and the casting expands. This expansion ensures that the casting fits the mold shape very well. The resultant microstructure of the iron matrix becomes pearlite. The microstructure of flaky graphite cast iron has sufficient strength for the shaft portion.

By contrast, the cam lobe needs high hardness to provide good wear resistance. If the rate of solidification of cast iron is fast, the included carbon forms into hard cementite ($Fe_3 C$). Iron combines with carbon to form cementite because graphite is difficult to nucleate at high solidification rates. Detailed explanations are given in Appendix C. Figure 5.11(b) shows the microstructure associated with rapid solidification. This microstructure is referred to as Ledeburite or chill, it is very hard and is highly suitable for the hardness requirements of cam lobes.

The cam lobe portion should be cooled rapidly in order to generate hard chill. An iron lump called a chiller is used for this purpose. The chiller is positioned at the cam lobe and takes heat away from the casting, giving a rapid solidification rate. The chiller is normally made of cast iron. Figure 5.10 illustrates the relative positioning of the chiller and cam. The chiller has a cam lobe-shaped cavity and is inserted into the sand mold prior to casting. Except for the chiller, the master mold consists of compacted sand. The shape and volume of the chiller determine how effective it is at absorbing heat, and it must be designed carefully to give the optimum cooling rate.

Figure 5.12 shows a section of a cam lobe, produced by etching the polished surface with dilute nitric acid. The pillar-like crystals, known as a columnar structure, align radially at the periphery, whilst they are not seen at the center. In Fig. 5.11(b), the columnar structure is aligned vertically, indicating that solidification advanced along the direction of heat flow. The crystal formation process during solidification was discussed more fully in Chapter 2.

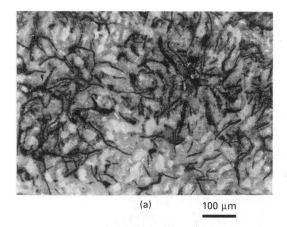

(a)   100 μm

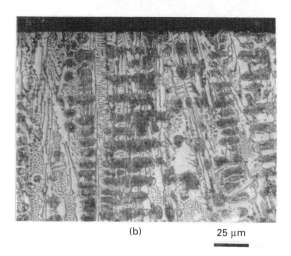

(b)   25 μm

5.11 (a) Flaky graphite of a shaft portion and (b) the chilled microstructure of a cam lobe. Chill has a mixed microstructure of cementite (white portion in (b)) and pearlite (gray portion). The hardness is around 50 HRC. Austenite and cementite appear simultaneously by eutectic solidification. The austenite portion transforms to pearlite during cooling. The eutectic solidification is called Ledeburite eutectic reaction. The additional quench-tempering changes pearlite to martensite. This heat treatment raises the hardness to 63 HRC.

Figure 5.13 shows the distribution of hardness at a cam lobe section, measured in three directions from the center to the periphery. The convex portion of the cam lobe shows a hardness of around 45 HRC, which is sufficient for this application, whilst the central portion is softer at 25 HRC. These microstructures correspond to the chill of Fig. 5.11(b) and the flaky graphite

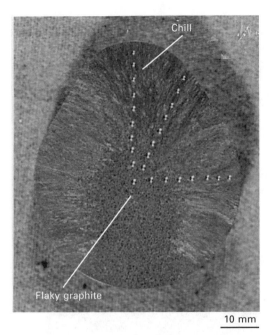

10 mm

5.12 Macrostructure of a cam lobe. The hardness measurement has indented small dents.

of Fig. 5.11(a), respectively. Generally, solidification starts from the surface, where the cooling speed is faster. Solidification in the central portion is slow due to the slow heat discharge rate, as confirmed by the hardness distribution.

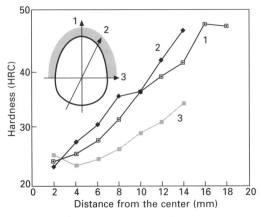

5.13 Hardness distribution of the cam lobe section. The surface shows a high hardness around 50 HRC because of rapid quenching by the chiller. The chiller has contacted the molten cast iron only at the gray part in the illustration.

It is not easy to produce hard chill without any graphite in mass production. A chill microstructure including graphite is soft and defective. The manufacturing process must control the chill hardness of the cam lobe to achieve the required value (45 HRC), while avoiding hard chill in the shaft portion. Hard chill in the shaft portion can break an expensive gun drill in subsequent machining operations, as described below. Generally, insufficient concentrations of C and Si are likely to cause chill even at slow solidification rates.

Inoculation[5] is a procedure aimed at solving the paradox that the hard chill and soft but strong shaft go together. As shown in Fig. 5.10, the inoculant is placed in the ladle before pouring. The inoculant adjusts the graphite shape (see Appendix D), preventing chill where it is not required. The inoculant is an alloy powder, such as Fe-Si, Ca-Si, or an alloy containing rare earth metals. The inoculation effect lasts for a limited period of time and gradually disappears after the inoculation (known as fading), thus it is important to time the inoculation accurately. The effect is similar to that of a nodularizer, as discussed in Chapter 2.

An alternative process to ensure hard chill at the cam lobes is remelting (see (2), Table 5.2). This process controls casting and chilling separately. The material is first cast to produce the flaky graphite microstructure. Then the surface of the cam lobe is partially remelted and then solidified rapidly to generate chill. Although it requires an additional process, this method provides better control of the hardness. However, if remelting is too slow, it can cause the shaft section to melt, so a high-energy heat source, such as a tungsten inert gas (TIG) torch, is used. The concentrated heat melts the surface of the cam lobe instantaneously.

5.3.2   Analysis of chemical composition of cast iron before pouring

Reusable raw materials, such as steel scraps from the body press process, are an abundant by-product of car manufacturing (Fig. 5.10). An electric furnace[6] melts the scrap with carbon (to raise carbon concentration) and ferro-alloys. The chemical composition must be checked before pouring. If molten cast iron has a high oxygen content, this will lower the strength of castings. Carbon and silicon[7] are used to remove oxygen from the melt, by reacting with oxygen to form CO_2 (which comes off as gas) or SiO_2 (which forms a glassy slag). In addition to this deoxidation effect, both elements greatly influence the strength of products through changing graphite shape and distribution.

Carbon concentration decreases rapidly in the melt, whereas silicon concentration does not. The concentration of other elements present, such as Mn, Cu, Ni, etc., does not change. This means that the carbon concentration in the melt must be analyzed and adjusted as necessary just before pouring. Analysis of the carbon concentration in the melt is based on a carbon equivalent (CE) value, obtained by measuring the solidification temperature of the cast iron. When molten cast iron is cooled, the gradient of the cooling curve becomes zero at the temperature at which solidification starts, due to the latent heat of solidification. A schematic example is shown in Fig. 5.14. Cooling continues when solidification is complete. The cooling curve of water shows similar behavior at the freezing point (0 °C), where ice and water coexist. If the water contains salt, the freezing temperature is lower, and solidification begins at a lower temperature. The same can be observed in the solidification behavior of cast iron.

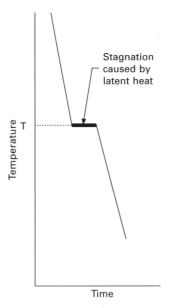

5.14 Cooling curve indicating temperature vs. time. A stagnation appears at the solidification point.

The solidification temperature of cast iron is proportional to the sum of the percentage of carbon and one-third of the percentage of silicon. This is known as the CE value and is measured using a CE meter. If the CE value is lower than that expected, the cast iron has a low concentration of carbon and silicon. A camshaft should chill only at the cam lobe, but if carbon and silicon concentrations are too low,[8] chill will be generated at the shaft portion as well. Measuring CE helps to reduce the risk of subsequent failure.

5.3.3   Finishing – boring and grinding

The camshaft needs a continuous, longitudinal central hole (Fig. 5.7) for the passage of oil. This also serves to reduce the weight. The hole is made using a gun drill (Fig. 5.15), which was originally developed for boring guns. The drill consists of a long pipe shaft with a cutting bit at the end. Machining oil is transmitted through the pipe to the bit during the drilling process. If hard chill has occurred in the central portion of the camshaft, this prevents the drill from boring effectively.

It is possible to eliminate the boring process by making the hole in the camshaft during the initial casting process. Figure 5.16 shows an example in cross-section. Excess metal is cut away using a long shell core.

75 mm

5.15 Gun drill. The right-hand end is a grip.

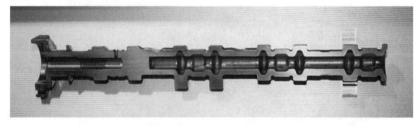

5.16 Chilled camshaft having a long hole as cast. To decrease the weight, the excess metal at the cam lobes is also removed.

The shape of the cam lobe has a direct influence on engine performance. A copy-grinding machine is used to finish the cam lobe. The grindstone traces a predetermined master cam. The hard chill means that each cam lobe has to be ground in small stages. Machine finishing is often followed by gas nitriding or manganese-phosphate conversion coating. These improve how the cam lobe adapts to the rocker arms during the running-in period.

As an alternative to chilled camshafts, a cam lobe with a microstructure of carbide and martensite (see (3), Table 5.2) has also been proposed. In this case, the camshaft is made from hardenable cast iron (Table 5.3). After machining, induction hardening on the cam lobe portion generates hard martensite, which gives a hardness of around 52 HRC. It has been reported that tough martensite is more resistant to pitting than the chill microstructure.

5.3.4   Composite structures

Camshafts can also be forged from Cr-Mo steel (Table 5.3). The entire camshaft is carburized and quench-tempered (see (4), Table 5.2). The multi-valve engine employs a greater number of valves, and the gap between these valves is consequently narrow, particularly in the small-bore-diameter engine, requiring short intervals between cam lobes. Chill hardening cannot be used where the gap between the cam lobes is narrow because of the difficulty in using the chiller, so forged camshafts are used.

Assembled camshafts (see (5), Table 5.2) consist of a hollow shaft and cam lobe pieces. Figure 5.17 gives an example. The cam lobe piece shown in Fig. 5.18 is made from a wear-resistant sintered material (Table 5.3) or hardened high carbon steel. The shaft portion is a steel tube.

5.17 Assembled camshaft using mechanical joining (hydroforming).

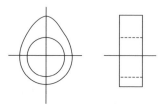

5.18 A cam lobe piece for assembled camshaft.

Figure 5.19 is a schematic representation of the powder metallurgy process used to shape and harden the cam lobe pieces for assembled camshafts. A mixture of powders that will produce the desired composition is prepared.

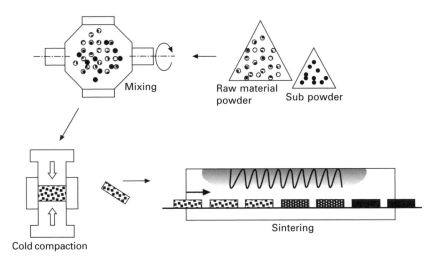

5.19 Powder metallurgical process.

This mixture is pressed into the die, which shapes the material in a process called cold compaction. The resulting shaped material is still porous and soft. The sintering process in the furnace removes pores through atomic diffusion and increases the density of the part. Generally, the compacted powder is heated to a temperature well below the melting point of the iron, usually between 1100 °C and 1250 °C, in continuous furnaces with a protective atmosphere. A density of 90% to 95% of the maximum theoretical value is quite normal, leaving between 5% to 10% porosity. This has some influence on the properties of the part, but the strength and hardness that can be achieved range from those of cast iron to those of hardened and tempered tool steel.

Sintering makes it possible to mechanically mix several dissimilar powders. Since sintering does not melt the powders, these can coexist in the sintered part so that the alloy composition can be very different from that produced during conventional solidification. A high amount of hard carbide with a fine dispersion, which is not possible in the normal casting process, is thus obtained and gives the cam lobes good wear resistance.

Powder metallurgy has the potential to produce near-net-shape parts, to permit a wide variety of alloy systems and to facilitate the manufacture of complex or unique shapes that would be impractical or impossible with other metalworking processes. In car engine parts, valve seats, main bearing caps and connecting rods (described in Chapter 9) are made by this process.

The chemical composition and hardness of the cam lobes can be adjusted in accordance with individual requirements. The alloy mixture for sintering contains small amounts of Cu. During sintering, the Cu melts and bonds the iron-alloy powder particles. The Cu works like a brazing filler metal, and the process is known as liquid phase sintering (as opposed to solid phase sintering, which does not generate a liquid phase).

In comparison with the chilled camshaft, the cost of the assembled camshaft is generally lower due to lower machining costs, and quality control is much better. Several bonding processes have been proposed for assembled camshafts.[9-11] Diffusion bonding, fusing or mechanical joining can all be used to join the cam lobes to the shaft. Diffusion bonding joins clean metal surfaces together through mutual diffusion when heated. Appendix I lists the various joining technologies.

Shave-joining[9] is a type of mechanical bonding. The surface of a steel tube is knurled to provide a rough surface. This steel tube is then inserted into the hole of the cam (Fig. 5.18). The rough surface produced fixes the cam lobe during fitting. More recently, a camshaft assembled by hydroforming[10] has been marketed (Fig. 5.16). In this process, a very high hydraulic pressure (around 100 MPa) swells the tube of the shaft from the inside (Fig. 5.20). The swelled shaft generates residual stresses in the cam lobe, which are sufficient to hold it in position. Hydroforming is used on some automotive suspension carriers and engine cradles because it can produce a complex twisted shape from a tube at low cost. Shrink fitting of the cam lobe piece to the steel tube has also been used.[11]

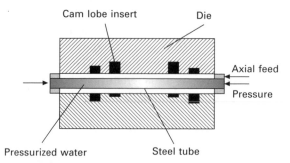

5.20 Assembling of camshaft using hydroforming process. The steel tube is placed in the die where the cam lobe inserts are already positioned. The internal pressurized water expands the steel tube to fix the cam lobe. The axial feeding pushes the end of the tube to minimize the wall thinning out.

5.4   Reducing friction in the valve train

The rotation of the cam lobe generates friction on the bucket tappet (Fig. 5.1) or rocker arm (Fig. 5.3). The bucket tappet drives the camshaft directly and is preferred for high-speed engines because it does not use an intermediate part (which reduces the rigidity of the valve train) and is of light weight. To reduce friction, a TiN PVD coating has been developed and marketed.[12]

The use of the rocker arm enclosing a roller bearing is becoming more common to reduce friction. Figure 5.21 shows a camshaft with the arm. The arm is shown in Fig. 5.22. It has been reported that the drive torque is as low as one-third that of the conventional rocker arm.

5.21 Camshaft and rocker arm type follower.

5.22 Rocker arm installing a roller (also called finger follower) for an OHC engine.

Friction is lower in the rocker arm with a roller bearing but the cam lobe receives high Hertzian stress. Figure 5.23[4] compares the pitting resistance of some materials used for cam lobes. The data were obtained using a wear-testing machine with an actual camshaft. In the comparison, sintered metal has the greatest durability at high contact pressure.

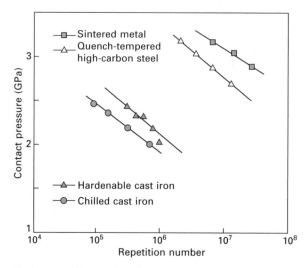

5.23 Comparison of pitting resistance measured by a camshaft and tappet wear testing machine using actual camshafts. The counterpart of the cam lobe is a roller made from a bearing steel (JIS-SUJ2). The rolling contact of the roller undergoes the contact pressure (vertical axis) with the repetition number (horizontal axis).

The rocker arm pictured in Fig. 5.22 is produced by investment casting of steel (described in Chapter 11). Rocker arms may also be produced using sheet metal forming, hot forging or sintering. The sintered arm[13] is injection-molded prior to sintering. Steel powder mixed with binder wax has sufficient viscosity to be injected into the metal die to shape a pre-form. After dewaxing, the pre-form is sintered in a vacuum furnace. This powder-metallurgical method is called metal injection molding (MIM).

5.5   Conclusions

A chilled camshaft has a composite structure which uses the metastable solidification of cast iron ingeniously. Chilled camshafts are used widely due to low costs. Cast iron is an excellent material because of its castability, but crystallized graphite lowers strength and can act as a point of weakness. Requirements for lightweight and strong materials have led to forged steels and aluminum alloys being substituted for cast iron.

5.6   References and notes

1. For example, let us imagine the flat bottom of a kettle on a table. The contact surface between the kettle and the table is a flat plane. The weight of the kettle is dispersed across the plane of contact. The contact pressure at the surface is determined by dividing the kettle weight by the contact area. By contrast, in the case of a kettle having a spherical bottom shape, although the unstable shape is fictitious, the contact portion becomes a geometric point. This applies when a heavy ball like a shot is placed on a table. The load concentrates at a point to cause an extremely high contact pressure. In practice the contact area takes a small circle because the two objects (shot and table) deform elastically. The concentrated stress is called Hertzian stress. It is possible to calculate the stress mathematically by considering the elastic deformation at the contact portion. H. Hertz is the scientist who first calculated it. The results have been given in various contact situations such as a sphere vs. a sphere, a plane vs. a sphere, etc.

2. *Automotive Handbook*: 5th edition, ed., by Bauer H., Warrendale, SAE Society of Automotive Engineers, (2000) 409.

3. Hoshi M. and Kobayashi M., *Kikaisekkei*, 24 (1980) 56, (in Japanese).

4. Quoted partially from E. Ogawa: *Tribologist*, 48(2003) 184.

5. A similar procedure called modification is also carried out to make the Si particle finer in high-Si aluminum alloys (Chapter 2).

6. Instead of an electric furnace, a cupola can be used to melt cast iron. It is difficult for the cupola to reach high temperatures (around 1400 °C) close to the melting temperature of pure iron. Since it cannot melt steel scraps, it uses pig iron produced by iron mills. The adjustment of chemical compositions is not easy in comparison with the electric furnace, yet its use is widespread because of the low energy cost.

7. Cast iron normally consists of a Fe-Si-C system. However, by substituting the Si with Al, Fe-Al-C system cast iron is possible. C. Defranq et al.: *Proceedings of the 40th International Conference*, Moscow, 3 (1973) 129.

8. Some minor elements such as trace S also influence chilling.

9. Egami Y. et al., *Soseitokakou*. 36 (1997) 941 (in Japanese). Nakamura Y. Egami Y. and Shimizu K., SAE Paper 960302.

10. Hydrodynamic technologies, Home page: http://www. hdt-gti.com, (2003).

11. Müller H. and Kaiser: A., SAE Paper 970001.

12. Masuda M., et al., SAE Paper 970002.

13. Nippon piston ring, Home page: http://www.npr.co.jp, (2003). Sokeizai, 1(2003) 31.

6

The valve and valve seat

6.1 Functions

Valves control the gas flowing into and out of the engine cylinder. The camshaft and valve spring make up the mechanism that lifts and closes the valves. The valve train determines the performance characteristics of four-stroke-cycle engines.

There are two types of valve, inlet and exhaust. Figure 6.1 shows an exhaust valve. An inlet valve has a similar form. The commonly used poppet valve[1] is mushroom-shaped. Figure 6.2 illustrates the parts of the valve. A

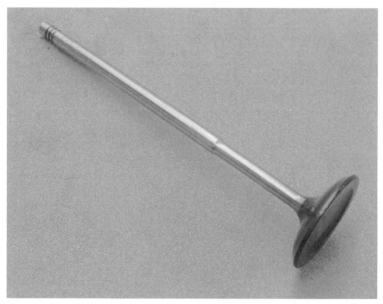

6.1 Exhaust valve. The inlet valve has a similar shape, but the crown size is normally larger than that of the exhaust valve.

cotter (not shown in Fig. 6.2) which fixes the valve spring retainer to the valve, is inserted into the cotter groove.

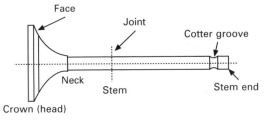

6.2 Nomenclatures of the valve. The shape from the crown to the neck is designed to give a smooth gas flow.

Figure 6.3 shows the position and relative motion of each part of the valve mechanism. The motion of the cam lobe drives the valve through the valve lifter. The valve spring pulls the valve back to its original position. During the compression stroke, the valve spring and combustion pressure help to ensure an air-tight seal between the valve and the valve seat.

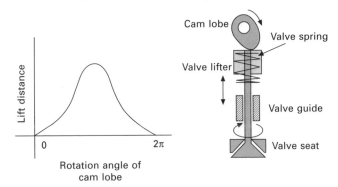

6.3 Rough sketch of a valve train showing valve lift distance in the valve timing diagram. The lift distance (vertical arrow) given by the cam lobe is the displacement along the axial direction of the valve.

One revolution of the camshaft gives the amount of valve lift shown in Fig. 6.3. The valve stem moves in the valve guide and also revolves slowly around the stem. The revolving torque is generated by the expansion and contraction of the valve spring.

An engine basically needs one inlet valve and one exhaust valve per cylinder but most modern engines use four valves per cylinder. This multi-valve configuration raises power output, because the increased inlet area gives a higher volume of gas flow. Contemporary five-valve engines use three inlet valves and two exhaust valves to increase trapping efficiency at medium revolutions.

Figure 6.4 summarizes the functions of the valve. The shape of the neck, from the crown to the valve stem, ensures that the gas runs smoothly. The valve typically receives an acceleration of 2000 m/s^2 under high temperatures. Valves must be of light weight to allow the rapid reciprocating motion.

In modern vehicles, various valve crown shapes are used. High-performance engines generally use recessed (vertical section is shown in Fig. 6.5) or tulip crown shapes (Fig. 6.13). The shape of the valve crown controls the flexibility of the valve face. Some high-speed engines need a flexible valve so that the valve does not bounce off its seat when closing. The recessed or tulip valve is elastically flexible as well as light.

Figure 6.5 illustrates typical temperature distributions[2] of an exhaust valve and an inlet valve. The combustion gas heats the inlet valve to around 400 °C, while the exhaust valve is heated to between 650 °C and 850 °C. The temperatures depend on engine types, with high-performance engines generating a great deal of heat. The exhaust valve always gets hotter than the inlet valve, because the cold inlet gas cools the inlet valve.

6.2   Alloy design of heat-resistant steels

Heat-resistant steels are classified into ferritic, martensitic and austenitic systems. The ferritic system is not suitable for engine valves because it is not strong enough at high operating temperatures.[3]

6.2.1   Martensitic steel

Since valves work under high-temperature (red heat) conditions, the materials need to be strong and corrosion-resistant at elevated temperatures. Most valves are made from heat-resistant stainless steels, which are resistant to sulfur corrosion as well as to oxidation at high temperatures. Table 6.1 lists the typical compositions of valve materials.

The inlet valve works at a temperature of approximately 400 °C, which is relatively low for iron-based materials. A martensitic heat-resistant steel such as JIS-SUH3 is commonly used. By contrast, the exhaust valve reaches approximately 850 °C at the valve crown, requiring an austenitic heat-resistant steel such as JIS-SUH35.

The martensitic system has a hard martensite microstructure. JIS-SUH3 is a typical alloy which gives superior wear resistance and intermediate temperature strength. Figure 6.6 shows the microstructure. The carbon content is around 0.4%, which raises hardness, and the alloyed Cr, Mo and Si give oxidation resistance. The cost of this type of alloy is relatively low.

Generally, material to be subjected to high operating temperatures must be tempered at a higher temperature than the actual working temperature. The tempered steel part is then stable when it is used at temperatures lower

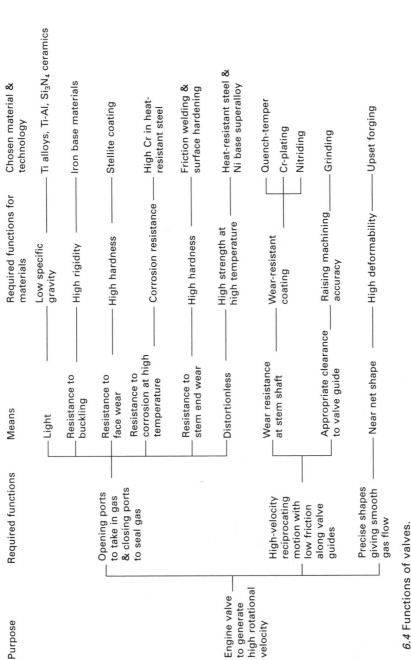

6.4 Functions of valves.

Purpose	Required functions	Means	Required functions for materials	Chosen material & technology
Engine valve to generate high rotational velocity	Opening ports to take in gas & closing ports to seal gas	Light	Low specific gravity	Ti alloys, Ti-Al, Si₃N₄ ceramics
		Resistance to buckling	High rigidity	Iron base materials
		Resistance to face wear	High hardness	Stellite coating
		Resistance to corrosion at high temperature	Corrosion resistance	High Cr in heat-resistant steel
		Resistance to stem end wear	High hardness	Friction welding & surface hardening
		Distortionless	High strength at high temperature	Heat-resistant steel & Ni base superalloy
	High-velocity reciprocating motion with low friction along valve guides	Wear resistance at stem shaft	Wear-resistant coating	Quench-temper / Cr-plating / Nitriding
		Appropriate clearance to valve guide	Raising machining accuracy	Grinding
	Precise shapes giving smooth gas flow	Near net shape	High deformability	Upset forging

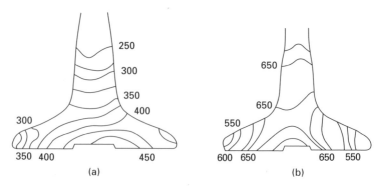

6.5 Temperature distribution (°C) of valves during operation. An air-cooled 200 cm³ engine. (a) Inlet valve (30 φmm). (b) Exhaust valve (26 φmm).

than the tempering temperature. The tempered part shows little microstructural change during operation because such changes have already occurred during the tempering process. Atomic diffusion controls microstructural change, and the higher the tempering temperature, the faster the change. It is important that the alloy composition of heat-resistant steel is designed so as not to lose strength in the tempering process. Valve steels consequently have high alloy concentrations because the alloying elements resist softening.

The martensitic alloy is quench-hardened, and the process consists of holding it at 1,000 °C followed by quenching, tempering at 750 °C and finally quenching in oil.[4] The temperature used for tempering martensitic heat-resistant steel is higher than that for normal carbon steel, because the microstructure is stable at high temperatures. The quenched steel softens more with higher tempering temperature (see Appendix F).

6.2.2   Austenitic steel

Figure 6.7 shows the microstructure of SUH 35. The fine dispersion of carbide and nitride in the stable austenitic matrix makes it strong at high temperatures. The high Cr and Ni concentrations (Table 6.1) make the austenitic matrix. These elements restrict A_1 transformation (the critical temperature in the transformation of steel that varies depending on the ratio of iron to other metals in the steel) to widen the austenitic area in the phase diagram (austenite decomposes into ferrite or pearlite at 723 °C, see Appendix C). As a result, a stable austenitic microstructure occurs even at room temperature. SUH35 keeps the austenite structure in the range from low to high temperature without causing martensitic transformation, therefore austenitic steel cannot be quench-hardened in the same way as martensitic steel.

The precipitated carbide in the austenitic heat-resistant steel increases creep resistance at high temperatures. For precipitation to occur, the following

Table 6.1 Chemical compositions of valve materials (%). The alloyed Cr forms a thick oxide film on the surface and prevents progressive corrosion at high temperatures. Since a high concentration of Cr does not make the alloy brittle, it is always alloyed in heat-resistant steels

Valve material	Name	C	Si	Mn	Ni	Cr	Mo	Fe	Others	Hardness	Heat treatment
Martensitic heat-resistant steel	JIS-SUH3	0.4	2	0.6	0.6	11	1	Balance	–	30 HRC	Quench & temper
Austenitic heat-resistant steel	JIS-SUH35	0.5	0.3	9	4	21	–	Balance	N: 0.5	35 HRC	Solution treatment & ageing
Co-base heat-resistant alloy	Stellite No. 6	1.2	1.1	0.5	3	28	1	3	Co: balance	57 HRC	Solution treatment & ageing

50 μm

6.6 Microstructures of JIS-SUH3, showing martensite with dispersed carbide.

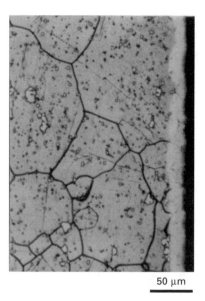

50 μm

6.7 Microstructure of JIS-SUH35 near the valve surface. Polygonal austenite grains with large carbides are observable. The nitride layer of 20 μm thick (white layer at the right edge) improves wear resistance.

three heat treatment stages must be followed: firstly, solution treatment at 1,100 °C, secondly, quenching, and finally, age hardening at 750 °C. When the alloying element, especially C, dissolves sufficiently during solution treatment, the fine carbide precipitates during ageing and increases high-temperature strength.

The strength depends on the environmental temperature, as shown in Fig. 6.8. In the low-temperature range below 500 °C, martensitic SUH3 is equal to or a little stronger than austenitic SUH 35. However, in the high-temperature range, SUH35 is stronger.

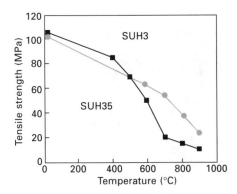

6.8 High-temperature strength of valve steels, SUH3 and SUH35. The martensitic SUH3 is stronger below 500 °C.

The reason that austenitic heat-resistant steel is stronger above 500 °C is due not only to the fine carbide dispersion, but also to the slow diffusion rates of elements in the austenite structure (FCC).[5] The slow diffusion rate of the included elements means that the microstructure generated by heat treatment hardly changes, thus maintaining strength at high temperatures.

Martensitic steel is hard below 500 °C, and is used in the mid-temperature range. By contrast, austenitic steel is used above 500 °C and is an appropriate choice where heat resistance is important.

6.3   The bonded valve using friction welding

Austenitic steel shows excellent strength at high temperatures, but, unlike martensitic steel, quench hardening is impossible due to the lack of martensitic transformation. Nitriding must be used as an additional heat treatment.

To obtain high wear resistance at the stem and stem end, martensitic steel is bonded to an austenitic steel crown. For this, friction welding is generally used. Figure 6.9 shows an as-bonded exhaust valve and Fig. 6.10 shows the microstructure at the weld joint.

Friction welding was first conducted successfully by A.I. Chudikov in 1954. Friction welding[6] is a method for producing welds whereby one part is rotated relative to, and in pressure contact with, another part to produce heat at the mating surfaces (Fig. 6.11). The friction generates the heat necessary

6.9 Friction-welded bond of an exhaust valve.

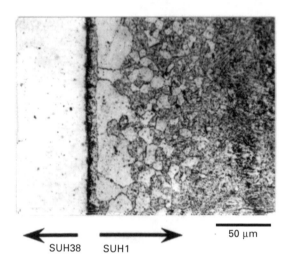

←———   ———→   50 μm
SUH38   SUH1

6.10 Microstructure of the bond between austenitic SUH38 and martensitic SUH1. Ferrite generated by the heat during friction welding appears in the SUH1 side. Solution treatment (heat treatment to dissolve solute atoms) was not carried out after the welding. The complete solution treatment and ageing can remove this ferrite.

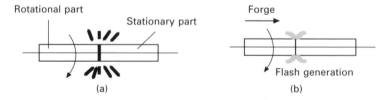

Rotational part   Forge
              Stationary part

        (a)   Flash generation
                                                      (b)

6.11 Schematic illustration of friction welding process. Welding is carried out in solid state without melting the materials. (a) The rotating rod (left) is slightly pressed to the stationary rod (right), so that friction heat is generated at the rubbing plane. (b) The heat softens the materials. Then the applied pressure along the axial direction welds the rods. At this forging stage, the oxide film at the rubbing plane discharges outside as flash and the resultant bond becomes clean. The flash is scraped off later.

for welding. One bar (the left portion in Fig. 6.11) rotates against the other, stationary bar under a small axial load for a given period. The friction heat generated makes the rubbing surfaces soft. As soon as rotation stops, the two parts are forged together. A butt joint is formed with strength close to the parent metals.

The joint portion does not melt, so the welding takes place in the solid phase. Since this mechanical solid phase process does not form macroscopic alloy phases at the bond, the joining of similar or some dissimilar materials is possible. For example, fused welding of aluminum with iron is generally impossible as the brittle Fe-Al compounds generated at the weld make the joint brittle. However, friction welding is possible because it does not form brittle compounds, and this method is typically used to combine carburized steel with stainless steel and to bond between two cast iron parts without generating brittle chill (Chapter 5). Friction welding is used only if one component can be rotated or moved linearly. A similar, solid-phase process is known as friction stir welding (FSW).[7] This method is used for butt-joining materials in plate form.

These mechanical, solid-phase welding processes give highly reliable joints with high productivity and low cost. A similar effect to friction welding can occur unintentionally as a result of adhesive wear, and this is termed seizure. Owing to its microstructure, the bond in the valve could be a source of weakness under lateral force, as shown in Fig. 6.10. The bonded portion is therefore usually located within the length of the valve guide. Various bonding technologies are used and are summarized in Appendix I.

Figure 6.12 illustrates the manufacturing process of a valve.[8] First, the sheared rod is friction-welded (process 3) and the part which will form the crown is made larger than the stem portion. To raise the material yield, upset forging is used to swell the crown portion from the stem diameter. The rod end is heated by resistance heating and upsetted (process 5). Die forging stamps the swollen portion into the crown shape (process 6) and the stem of the bonded valve is heated and quench hardened (process 18).

Exhaust valves reach very high temperatures and their strength at such temperatures relies on selecting a suitable material. However, there is also a way to control the temperature of the valve structurally, by using a hollow valve containing sodium. The Na in the stem melts during operation and the liquid metal carries heat from the crown to the stem. Na is solid at room temperature, but melts at 98 °C and the valve stem works as a heat pipe. Reciprocating aeroplane engines used this technique during the Second World War, as do high-power-output car engines at present. Historically, a valve enclosing a liquid such as water or mercury was first tried in the UK in 1925,[9] and also tried with a fused salt, KNO_3 or $NaNO_3$, in the USA.

Friction welding is used to enclose Na in the valve stem. In the friction-welded valve (Fig 6.9), the crown side is first drilled to make a cavity for the

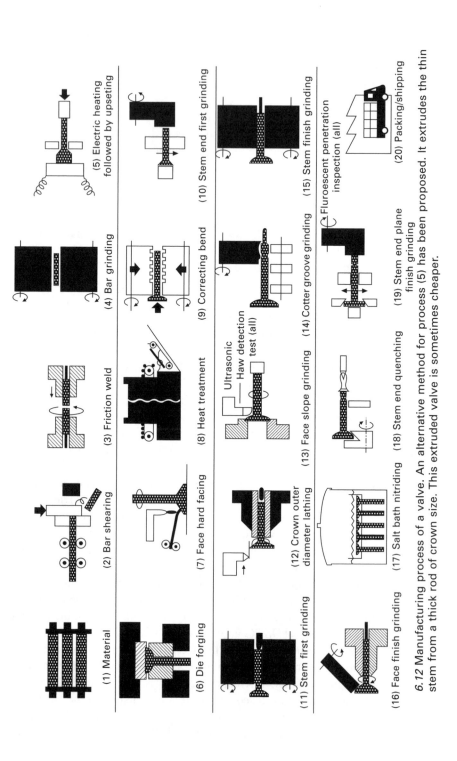

(1) Material

(2) Bar shearing

(3) Friction weld

(4) Bar grinding

(5) Electric heating followed by upseting

(6) Die forging

(7) Face hard facing

(8) Heat treatment

(9) Correcting bend

(10) Stem end first grinding

(11) Stem first grinding

(12) Crown outer diameter lathing

(13) Face slope grinding — Ultrasonic Haw detection test (all)

(14) Cotter groove grinding

(15) Stem finish grinding

(16) Face finish grinding

(17) Salt bath nitriding

(18) Stem end quenching

(19) Stem end plane finish grinding

(20) Packing/shipping — Fluroescent penetration inspection (all)

6.12 Manufacturing process of a valve. An alternative method for process (5) has been proposed. It extrudes the thin stem from a thick rod of crown size. This extruded valve is sometimes cheaper.

Na. Na and nitrogen are then placed in the hole and the crown side is friction-welded to the shaft.

6.4   Increasing wear resistance

6.4.1   Stellite coating

The carbon soot formed by combustion can stick to the valve, hindering valve closure and consequently causing leakage. To prevent this, the valve revolves during reciprocative motion, as described earlier (Fig. 6.3). The rotation rubs off the soot and prevents uneven wear of the valve face and seat. The face is exposed to high-temperature combustion gas and so this rubbing occurs without oil lubrication. The valve material itself does not have high wear resistance, so must be hardened to improve wear resistance at high temperatures.

Wear resistance in the valve face is improved by a process known as hard facing (process 7 in Fig. 6.12). The valve face is gradually coated with melted stellite powder, a cobalt-based heat-resistant alloy, until the entire circumference is overlaid. A plasma welder[10] or a gas welder is used to melt the powder. Figure 6.13(a) shows a cross-section of an exhaust valve crown. The microstructure of the stellite is a typical dendrite, characteristic of cast microstructures. The result is a hardness value of around 57 HRC.

Table 6.1 gives the chemical composition of stellite. Cobalt-based heat-resistant alloys have excellent heat resistance compared to Fe or Ni-based alloys but are costly. Hence, a small amount is used only where their high heat-resistant properties are required. Among stellite alloys, there are alloys with increased Ni and W, which are much more wear resistant. Recently, instead of stellite, Fe-based hard facing materials[11] have been developed to reduce costs. The typical composition is Fe-1.8%C-12Mn-20Ni-20Cr-10Mo.

Wear in the valve lifter results from contact with the valve stem end (Fig. 6.2 right end; valve stem end). The valve stem end is also coated with stellite to increase wear resistance as a substitute for quench hardening (process 18 in Fig. 6.12).

The valve stem also rubs against the inside of the valve guide. To improve wear resistance here, salt bath nitriding (process 17 in Fig. 6.12) or hard chromium plating are used. Salt bath nitriding is preferred for high-chromium heat-resistant steel (see Appendix H), and can produce a more homogeneous nitrided layer compared to gas nitriding.[12]

6.4.2   The Ni-based superalloy valve

Stellite is expensive to use. Valves that use Ni-based superalloys, such as Inconel 751[13] or Nimonic 80A, have been developed as an alternative to hard

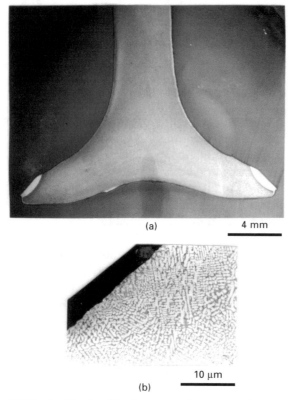

(a)   4 mm

(b)   10 μm

6.13 (a) Stellite hard facing at a valve crown. (b) Magnified view of the stellite microstructure.

facing. Valves without a stellite coating are becoming increasingly common as exhaust valves in high-output engines. Table 6.2 shows the chemical compositions of Inconel 751 and Nimonic 80A. Both are stronger at high temperatures than austenitic heat-resistant steel.

Ni-based superalloys get their increased strength due to precipitation hardening. The hardening mechanism is the same as for austenitic valve

Table 6.2 Ni-base valve material compositions (%). There are much stronger materials in Ni-base superalloys. However, these are cast alloys and are impossible to shape by forging

Ni base superalloy	C	Si	Mn	Ni	Cr	Co	Ti	Al	Fe	Nb+Ta	Hardness
Inconel 751	0.1	0.5	1.0	Balance	15.0	–	2.5	1.0	7.0	1.0	38 HRC
Nimonic 80A	0.1	1.0	1.0	Balance	20.0	2.0	2.5	1.7	5.0	–	32 HRC

steels, microscopically, the mechanism is similar to the age hardening of piston alloys (Chapter 3). Coherent precipitation gives high strength by raising the internal stress of the matrix. In the Ni-based superalloy, the high temperature strength is at a maximum when a coherent precipitate Ni_3 (AlTi) appears (see Appendix G).

Ni-based superalloys make the valve face strong to remove the need for stellite, but cannot give enough wear resistance at the stem or stem end. Nitriding is not possible for Ni-based superalloys due to the material properties of Ni, which are similar to austenitic stainless steel. To overcome this, a small piece of martensitic steel is friction-welded to the valve stem end.

6.5   Lighter valves using other materials

6.5.1   Ceramics

New materials for producing lightweight valves have been tested. For engines with large diameter valves, lightweight materials are a definite advantage. Silicon nitride (Si_3N_4) valves, shown in Fig. 6.14, have been researched extensively. Si_3N_4 weighs as little as 3.2 g/cm^3. It has a bending strength of 970 MPa at room temperature and 890 MPa even at 800 °C. By contrast, the austenitic steel SUH35 shows a bending strength of only 400 MPa at 800 °C (Fig. 6.8). It has been reported that the weight reduction from using Si_3N_4 instead of a heat-resistant steel valve is 40%.[14]

Ceramic materials are brittle under tensile stress conditions, so design and material quality are very important. Figure 6.15 shows the manufacturing process. Silicon nitride powder is first molded and then baked. To increase reliability, particular attention is paid to the purity of the materials, grain size and the baking process.

Some ceramic parts have already been marketed as engine parts. These include insulators for ignition plugs, the honeycomb for exhaust gas converters, turbo-charger rotors, wear-resistant chips in a valve rocker arm, and the pre-chamber for diesel engines. However, despite vigorous research efforts, ceramic valves have not yet been marketed.

6.5.2   Titanium alloys

Titanium alloys have also been used for valves. The Toyota motor company marketed an exhaust valve in 1998 made from a Ti matrix composite alloy, Ti-6%Al-4Sn-4Zr-1Nb-1Mo-0.2Si-0.3O, containing TiB particles (5% by volume).[15] The relative weight was about 40% lower, which also enabled a 16% decrease in valve spring weight. It was reported that a 10% increase in maximum rotational velocity and a 20% reduction in friction were obtained.

6.14 Si$_3$N$_4$ ceramics valve (courtesy of NGK Insulators, Ltd.).

Powder-metallurgy is the process used to produce an extruded bar for hot forging. This is similar to the process for the PM cylinder liner (Chapter 2) and piston alloy (Chapter 3). A mixture of TiH$_2$, TiB$_2$, and Al-25%Sn-25Zr-6Nb-6Mo-1.2Si powders is sintered at high temperatures. During this sintering, densification through diffusion takes place and the chemical reaction forms TiB particles. This process is called *in-situ* reactive combustion synthesis. The sintered material is extruded into a bar, which is then forged into a valve using the same process as that used for steel valves. Additional surface treatments are not necessary because of the high wear resistance of this composite. Appendix L summarizes the metal matrix composites in engines.

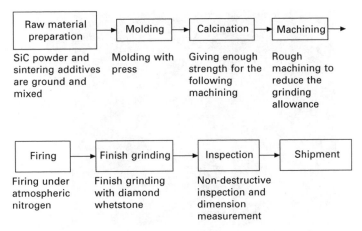

6.15 Production process of a silicon nitride ceramics valve.

Another Ti exhaust valve has also been marketed.[16] This valve is not manufactured using powder-metallurgy, but instead uses cast and rolled Ti-6%Al-2Sn-4Zr2Mo-Si alloy, which is widely found in the compressor disk of jet engines. It has a dual structure, where the crown portion has an acicular microstructure and the stem portion an equiaxed one. Figure 6.16 shows these microstructures. The acicular microstructure is stronger than the equiaxed one above 600 °C, and is generated by upset forging of the crown portion above the β-transus temperature (995 °C). Plasma carburizing is used to increase wear resistance.

A Ti inlet valve can also reduce weight. Since inlet valves do not require the same high heat resistance properties as exhaust valves, normally Ti-6%Al-4V alloy is used. Exhaust valves made from a Ti-Al intermetallic compound[17,18] have also been investigated but are not yet commercially available. The application of Ti alloys for automotive use is summarized in references 19–21.

6.6   The valve seat

The valve seat insert has a cone-shaped surface as shown in Fig. 6.17. The seat is pressed into the aluminum cylinder head (see Chapter 7) and seals in combustion gas, so needs to have good wear resistance to ensure an accurate and air-tight seal. Since heat escapes through the cylinder head, the operating temperature for the seat will be lower than that of the valve.

Table 6.3 lists typical chemical compositions of valve seats. In the past, the lead additives in fuel lubricated the contact points between the valve and valve seat, since lead acts as a solid lubricant at high temperatures. However, unleaded fuel by its very nature does not contain lead-type lubricants. When

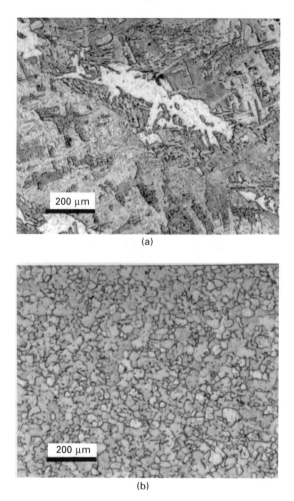

6.16 Microstructures of a Ti valve; (a) acicular microstructure at the crown and (b) equiaxed one at the stem.

leaded petrol was replaced with unleaded alternatives, valve seat materials had to be developed to cope with the changed lubrication conditions.

In the past, valve seats were manufactured from cast iron, but now sintered materials are more common. Figure 6.18 shows the microstructure of a valve seat material. Generally, valve seat materials are iron-based sintered alloys containing increased Ni, Co, Cr and W. The high Cr and W compositions increase carbide dispersion. The exhaust valve seat contains the highest levels because it is exposed to more severe wear at higher temperatures. Cu and/or Pb[22] are included as solid lubricants.

6.17 Valve seat inserts for inlet (right) and exhaust (left).

Table 6.3 Valve seat material compositions (%)

Valve seat material	C	Ni	Cr	Mo	Cu	W	Co	Fe	Hardness	Heat treatment
Exhaust	1.5	2.0	8.0	0.8	18.0	2.0	8.0	Balance	35 HRC	Quench & temper
Inlet	1.5	–	0.5	–	4.0	–	–	Balance	100 HRB	Quench & temper

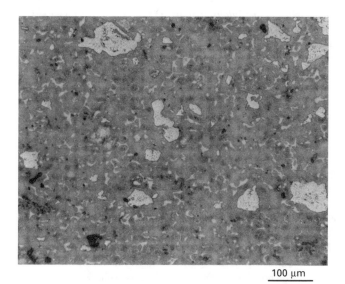

100 μm

6.18 Microstructure of a valve seat material dispersing large globular W, V and/or Cr carbides around 30 μm (about 1700 HV). The matrix shows sorbite microstructure (about 300 HV). The infiltrated Cu is also observable among steel particles. The steel particles are sintered first. It contains pores among the particles. The Cu is infiltrated into the pores.

6.7   Conclusions

The exhaust valve, exhaust pipe, exhaust gas turbine in a turbo-charger, honeycomb catalyst holder and brake disks are exposed to high operating temperatures of around 900 °C. The exhaust valve, exhaust gas turbine and honeycomb always operate under red-hot conditions. For these parts, iron-based heat-resistant alloys, nickel-based superalloys and ceramics are functionally competitive.

The exhaust valve seat, brake pad and friction plate (for a dry clutch), do not receive lubricating oil during operation, so these operate in the tribology area, where composite materials are most suitable.

6.8   References and notes

1. Iwata T., *Nainenkikan*, 4 (1965) 57 (in Japanese).
2. The data is measured using a temperature-measuring valve. This experimental valve is first made from a material (for example, JIS-SUJ2) showing temper softening during operation. After the engine operation, the decreased hardness of the valve can be used to estimate the temperature with reference to the master curve. It is similar to the method to estimate the piston temperature (Chapter 3).

   In addition, the hardness change of the electrode metal of a plug can be used to evaluate the combustion state in the combustion chamber. The hardness change of the parts exposed to heat can be used to estimate the operating temperature. It is simple and convenient in performance development. The following is a general review. Asakura S. *et al.*: *Jidoushagijutu*, 33 (1979) 775 (in Japanese).
3. Ferrite steels containing high Cr and low C do not show transformation up to high temperatures. However, ferritic steels are not used under high stress, because the creep strength rapidly decreases above 500 °C.
4. To avoid temper embrittlement, cooling should be rapid over the temperature range from 350 to 550 °C.
5. Tsuda M. and Nemoto R., *The 5th Nishiyama Kinen Gijutsu Kouza*, Nihon Tekkoukyoukai, (1994) 135 (in Japanese).
6. *Friction welding:* Corona Publishing, Tokyo, (1979) (in Japanese).
7. Dawes C.J., *Weld Met. Fabr.*, 63(1995)13.
8. Nittan Valve Co., Ltd. Company guide, (1997).
9. Tomituka K., *Nainenkikannorekishi*, Sanei Publishing, (1987), 104 (in Japanese).
10. Takeuchi H., *et al.*: *Yousetsu Gijutsu*, September, (1985) 20 (in Japanese).
11. FUJI OOZX Inc. catalogue, (2000).
12. The recently modified gas nitriding can give a homogeneous nitrided layer.
13. Ni is expensive. An engine valve without Ni has been developed. Sato K. *et al. Honda R & D Technical Review*, 9 (1997) 185 (in Japanese).
14. Moergenthaler K., *Proceedings of the 6th International Symposium on Ceramic Materials and Components for Engines.* Edited by Niihara K. *et al.*, (1997) 46.
15. Yamaguchi T., *et al.*, SAE Paper 2000-01-0905.
16. Mouri A., *et al. Titan*, 50 (2002) 45 (in Japanese).
17. Maki K., *et al.* SAE Paper 96030.
18. Blum M., *et al.*, *Mater. Sci. Eng.*, A329-331 (2002) 616.

19. Takayama I. and Yamazaki T., *Shinnitetu Gihou*, 375 (2001) 118 (in Japanese).
20. Yamashita Y., *et al.,* Nippon Steel Technical report, 85 (2002) 11.
21. Fujii H., Takahashi K. and Yamashita Y., *Shinnitetu Gihou*, 378 (2003) 62 (in Japanese).
22. The lead in the valve seat material was first added by aiming at a similar effect to leaded gasoline. Initially, a lead content of about 4% was used. Kawakita T., *et al.,* *Proceedings of 1973 International Powder Metallurgy Conference*, eds Hauser H.H. and Smith W.E., Metal Powder Industries Federation and American Powder Metallurgy Institute. Recently, the Cu or Pb content or the addition itself tends to decrease.

7

The valve spring

7.1   Functions

Figure 7.1 shows a valve spring. The valve spring is a helical spring used to close the poppet valve and maintain an air-tight seal by forcing the valve to

7.1 Valve spring. Generally, coil springs of a wire diameter below 5 mm ϕ are cold-formed at room temperature, while wires above 11 mm are normally hot-formed. Compression valve springs are provided with the ends plain and ground.

the valve seat. A spring accumulates kinetic energy during contraction and the energy is dissipated upon expansion. There are many types, shapes and sizes of steel springs.

The valve train consists mainly of valves, valve springs and camshafts. At low camshaft revolutions, the valve spring can follow the valve lift easily so that the valve moves regularly. By contrast, at high revolutions, it is more difficult for the valve and valve spring to follow the cam. Valve float is the term given to unwanted movements of the valve and valve spring due to their inertial weights. To avoid this, the load of the valve spring should be set high. The load applied at the longest length is called the set load, and the valve spring is always set to have a high compressive stress above set load conditions. Figure 7.2 shows double springs, which are used to raise the set load while minimizing the increase in height.

7.2 Double springs installed in a bucket type valve lifter.

Another resistance phenomenon that occurs at high revolutions is surging, due to resonance. Surging occurs when each turn of the coil spring vibrates up and down at high frequency, independently of the motion of the entire spring. It takes place when the natural frequency of the valve spring coincides with the particular rotational speed of the engine. Generally, surging occurs at high revolutions, and the surging stress generated is superimposed on the normal stress. The total stress is likely to exceed the allowable fatigue limit of the spring material and can break the spring. A variable pitch spring reduces the risk of surging. This spring has two portions along the length, a roughly coiled portion and a densely coiled portion, which ensures that the natural frequency of the spring is not constant and therefore not susceptible to resonance.

7.2   Steel wires

The valve spring should be made as light as possible using a thin wire with a high spring limit value. High straining above the elastic limit (yield stress) causes plastic deformation of the spring. A light spring relies on the material property of the wire, and there are two possible methods of achieving a light spring. Firstly, a material with a high Young's modulus can be used. This ensures an adequate spring constant even if the wire is thin. The alternative is to raise the yield point (Appendix K) of the material to prevent yielding even at high stress. This enables the spring to withstand high loading and deflection. The elastic modulus of iron is around 21 GPa and is not changed by heat treatment, therefore, only the second method can be used to create a light spring.

It is preferable to use material at just below its yield stress, but this situation is likely to cause fatigue failure. In addition, despite oil cooling, the valve spring becomes heated in the cylinder head. Consequently, the spring material needs good formability, shape stability during operation, heat-resistance at the engine oil temperature and high fatigue strength.

To meet these demands, an oil-tempered wire (JIS SWOSC-V[1]) made of Si-Cr steel is generally used for valve springs. Table 7.1 shows some typical chemical compositions. It is a high-carbon steel containing raised levels of Si and Cr. Figure 7.3 shows the microstructure of the steel. It is normally

Table 7.1 Chemical composition of valve spring material (%). The quantities of Si and Cr are high

Valve spring material (JIS)	C	Si	Mn	Cr	V	Ni	Tensile strength (GPa)	Heat treatment	Remarks
SWOSC-V	0.55	1.45	0.7	0.7	–	–	1.9	Quench-temper	SAE 9254
High strength oil-tem-pered wire	0.59	1.95	0.85	0.9	0.1	0.25	2.05	Quench-temper	–
SWP-V (piano wire)	0.82	0.25	0.5	–	–	–	1.6	Patenting	SAE 1080
High-Si piano wire	0.82	0.93	0.75	–	–	–	1.9	Patenting	–

25 mm

7.3 Microstructure of a valve spring material. Currently, failure caused by nonmetallic inclusions is rare. The allowable inclusion size should be below 20 μm. Yet, it is harmful when the inclusion is at the spring surface.

used in the martensitic form, generated by quenching and tempering. The tensile strength measures up to 1.7 GPa. There is also a higher-strength wire containing slightly higher alloying elements (listed in the table) than SWOSC-V, but a strength of around 2.4 GPa is thought to be an upper limit for valve spring material.

Applying the heat treatment known as oil-tempering (oil quenching and tempering) before coiling gives the wire sufficient elastic properties. The wire is quenched into oil from the austenite temperature followed by tempering at 320 to 400 °C, in a continuous process. Included Si strengthens the ferrite matrix during this heat treatment. It is difficult to control high Si quantities precisely in the steel-making process. This manufacturing limitation results in typical sizes of the oil-tempered wire being in the range of 0.5 to 8 mm diameter. During tempering at around 350 °C, carbon atoms or carbide immobilize the high-density dislocation introduced during quenching, which gives extremely high strength.

Normally, tempering at around 350 °C causes an adverse effect in high carbon steel, making it very brittle. This is known as low-temperature temper-embrittlement (see Appendix F), and thus tempering at around 350 °C should normally be avoided. In Si-Cr steels, however, the alloyed Si raises the temperature at which embrittlement occurs, enabling low-temperature tempering and ensuring that the steel is extremely strong. For this reason, the Si content is increased to as much as 1.4%.

Oil-tempering is carried out on a straight wire to prevent the valve spring from retaining stress and avoiding other unfavorable conditions, such as crook, but the lack of fiber texture (described in Chapter 8) in the microstructure makes the oil-tempered wire fragile under sharp bending conditions.

Before 1940, only piano wires were used, and although piano wires are still used because of their low cost, the proportion found nowadays in automotive valve springs is estimated to be below 5%. The piano wire listed in Table 7.1 is a newly developed high-Si type[2] with increased Si and Mn content. The fatigue strength is a little better than that of SWOSC-V and the resistance to sag is increased by hot-setting (described below).

Steels for cold-wound springs differ from other constructional steels chiefly in the degree of cold work, the higher carbon content, the fact that they can be supplied in the pre-tempered condition and their higher surface quality. The fatigue strength of spring steel is very sensitive to defects such as surface scratches, decarburization and nonmetallic inclusions inside the steel. Decarburization occurs when the carbon content at the surface is reduced during heat treatment.

Nonmetallic inclusions in spring steel initiate fatigue failure, and are typically hard nonmetallic particles of Al_2O_3, (MnO, MgO or CaO)-Al_2O_3 and SiO_2 generated in the steel-making processes. It has been determined that the performance of steel wires having a tensile strength above 1.8 GPa[3] is largely influenced by these inclusions. Sensitivity to defects increases rapidly in the high strength region. Secondary refining technologies (see Chapter 9) have greatly reduced the amount of nonmetallic inclusions and have lengthened the fatigue life of steel.

7.3   Coiling a spring

A four-cylinder engine with five valves per cylinder uses forty valve springs. If the properties of each spring are different, stable engine operation is not possible. Variations of quality should be controlled to within a suitable range.

The oil-tempered wire is coiled at room temperature to form the spring. The valve spring is a cold-wound spring (Fig. 7.4 illustrates the manufacturing process),[4] and the process is as follows:

1.   The wire is continuously coiled to spring one coil at a time.
2.   The spring is annealed at around 400 °C to remove strain.
3.   A double-headed surface-grinding machine makes both ends of the spring parallel.
4.   The spring is shot peened (described below) thoroughly under rotation, to raise fatigue strength.
5.   The spring is annealed again for a short period at around 250 °C to stabilize the strain caused by shot peening, and for heavy-duty use, an

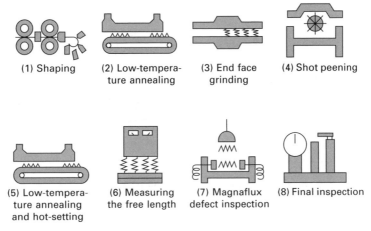

(1) Shaping   (2) Low-tempera-   (3) End face   (4) Shot peening
                   ture annealing   grinding

(5) Low-tempera-   (6) Measuring   (7) Magnaflux   (8) Final inspection
    ture annealing   the free length   defect inspection
    and hot-setting

7.4 Manufacturing process of valve springs.

additional process called hot setting[5] is implemented. Hot setting improves the load stability of the spring at engine temperature. The process consists of low-temperature annealing followed by water-cooling under a slight compression. By subjecting the spring to this process, primary creep is removed from the material and subsequent load loss in service is minimized.

6.   The spring is finally inspected for shipping.

If a spring is not manufactured properly, it will sag. Sag appears when the load which the spring should bear decreases during use. Oil-tempered wire already has the required yield strength and tensile strength before coiling, but the additional straining in the coiling process lowers resistance to sag. To decrease sag, a process called pre-setting or setting is carried out in the final stage of the manufacturing process. Setting intentionally gives deflection with a slight plastic deformation. Figure 7.5 illustrates the principle using a load-deflection curve. The spring material has the yield point A before setting. The setting loads and strains the spring material up to B. After setting, the load reduces down to X but the strain remains as the plastic deformation OX. In reloading after setting, the material yields at point B. This means that the yield point has increased from A to B. The additional low-temperature annealing at 150–300 °C increases yield point to C, removing the microyielding phenomenon.

Sag is caused by the plastic deformation of the spring at low stress within the macroscopic elastic limit. The dislocations introduced during cold working stay at unstable positions. These dislocations can drift even at low stresses, below the macroscopic yield stress of the material. If a load is applied under these conditions, the spring demonstrates microyielding (a small plastic deformation at low stress). The additional low temperature annealing immobilizes the free dislocations with C, carbide or N. This prevents

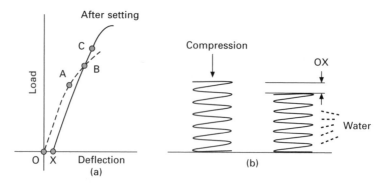

7.5 The principle of setting. (a) The deflection of a spring by setting.
(b) Increase in yield stress by setting. OAB shows load-deflection
before setting, while XBC after setting. OX is a deflection given by
setting.

microyielding and raises the yield point to C. Hot setting is the process that
allows both setting and low-temperature annealing to occur at the same time.
It is carried out in a furnace at temperatures of 200–400 °C. Hot setting
therefore raises the spring limit value[6] and improves sag resistance, and is
now widely used in the manufacturing process of springs.

A valve spring experiences higher stress on the inside of the coil than on
the outside during operation. Fatigue failure therefore usually begins on the
inside. Figure 7.6 is a typical example. The failure originated at the scratch
inside the coil. To prevent this, a wire with a non-circular cross-section (e.g.,
elliptical or unsymmetric clothoid) is sometimes used.

7.4   Improving fatigue strength by shot peening

The valve spring works under high stress. The allowable stress on a valve
spring has risen from 800 MPa in 1975 to over 1,300 MPa today. To increase
fatigue strength, a surface treatment called shot peening[7] is carried out (process
4 in Figure 7.4). Shot peening subjects the surface to vigorous bombardment
with small steel balls (grains) using air injection. The impacts of the balls
spread the surface plastically, while having no effect on the material beneath
the surface. Shot peening generates compressive residual stress in the surface
layer. During operation, tensile stress appears in the spring surface, but the
residual compression stress[8] effectively counteracts the tensile stress and
increases the fatigue life tenfold.[9]

Figure 7.7 is an example of a residual stress distribution resulting from
shot peening for about 45 minutes, as measured by X-ray.[10] The negative
sign indicates that the stress is compressive. The stress is indicated by the
full line in the very shallow portion at the surface. The compressive residual
stress reaches a maximum value at 0.1 mm below the surface. The stress

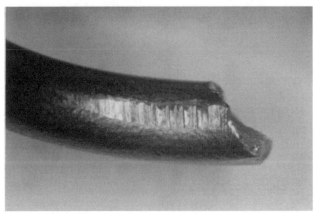

7.6 Shear fracture caused by twist fatigue. The crack initiated at a scratch inside the coil turning. The lower photo observes the fracture from inside. Vertical scratches are observable.

valley just below the surface is caused by the fact that the increased surface temperature during shot peening relaxed and decreased the compressive stress. Such a distribution appears when single-size balls are used. Additional shot peening using smaller balls eliminates the decline in residual stress, giving an ideal residual stress distribution, as indicated by the broken line shown in Fig. 7.7. This is known as double shot peening.

Figure 7.8 shows the effect of residual stress. The cross-section of a part with parallel surfaces is shown. Under the applied tensile stress D, the tensile stress at the surface is reduced to F from D by the residual compressive stress A. Generally, fatigue cracking is likely to initiate at the surface where the tensile stress is at a maximum. Hence, it is very important to control the

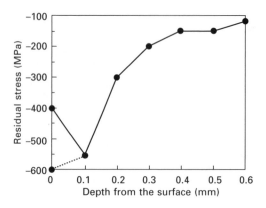

7.7 Residual stress distribution measured by X-ray after shot peening. The double shot peening modifies the surface stress (broken line).

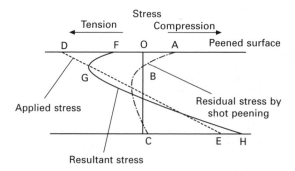

7.8 Effect of residual stress on the resultant stress.

distribution and intensity of such compressive residual stress in making valve springs. When steels containing high Si content are heat treated, decarburization is likely to occur at the surface. The stress introduced by shot peening can compensate for the adverse effects of decarburization. Shot peening is also used to prevent stress corrosion cracking, wear, and so on.

For much greater strength, it has been suggested that valve springs should be nitrided before shot peening. This can give a tensile strength of 2.2 GPa or more. The high-strength steel[3] in Table 7.1, which contains higher C, Si, Cr together with V and Ni, can be used in this way. Since the higher Si, Cr and V content restrict softening during tempering, the decrease in hardness during nitriding is less. The steel, therefore, retains a high fatigue resistance without affecting sag resistance. This type of spring is used in high-output engines.

7.5   The cylinder head

Figure 7.9 shows a typical cylinder head. Together with the piston, the cylinder head provides the desired shape of combustion chamber. In two-stroke engines, the function is limited to this. By contrast, in the vast majority of four-stroke engines, the cylinder head mounts the entire valve gear and is a basic framework for housing the gas-exchange valves as well as the spark plugs and injectors. Figure 7.10 is a view of a cylinder head with five valves per combustion dome.

7.9 Cylinder head.

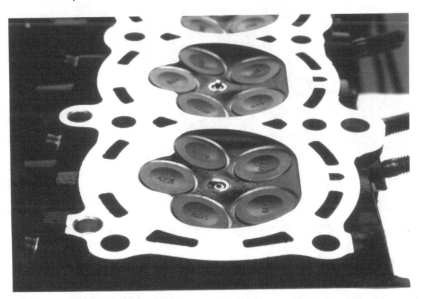

7.10 Cylinder head observed from the combustion chamber side.

In trucks and large industrial engines, individual cylinder heads are often used on each cylinder for better sealing force distribution and easier maintenance and repair. In car engines, one cylinder head is usually employed for all cylinders together. The cylinder heads on water-cooled diesel truck engines are usually made of cast iron. By contrast, all petrol and diesel engines for cars use aluminum cylinder heads due to the superior heat dissipation and lower weight. In cars, the cylinder head is normally made of aluminum even when the cylinder block is cast iron.

The three fluids, combustion gas, coolant and lubricating oil, flow independently in the cylinder head. Figure 7.11 shows a model of the coolant and gas circuits in a cylinder head. These circuits follow complex three-dimensional routes, so cylinder heads are generally produced by casting. Gravity casting or low-pressure casting using sand molds or metal dies are used. These circuits are cast hollow by using sand cores installed in the holder mold. High-pressure die casting is not used because the sand cores are fragile and cannot endure the high injection pressure of molten aluminum.

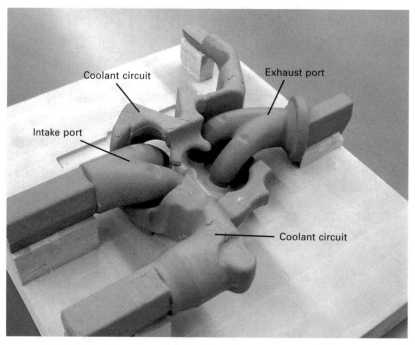

7.11 Gas and coolant circuits in a cylinder head.

The cylinder head uses JIS-AC4B, an Al-Si-Cu system alloy (see Table 7.2). This is a typical alloy for gravity and low-pressure casting and is widely used in various fields. In order to increase toughness, the Si content is decreased

Table 7.2 Chemical composition of cylinder head material (%)

JIS	Si	Fe	Cu	Mn	Mg	Zn	Ni	Ti	Pb	Sn	Cr	Al	Hard-ness	Heat treatment
AC4B	8.0	1.0	3.0	0.5	0.5	1.0	0.4	0.2	0.2	0.1	0.2	Balance	75 HB	T6

to 8.0% and an age-hardening effect is given by adding 3% Cu. The alloy has good castability as well as high strength at intermediate temperature range. The tensile strength is 111–176 MPa in the as-cast state and 218–299 MPa after T6 heat treatment.

The cylinder head receives a great amount of heat from the cylinder block, so dimensional stability is required over a long period of time. Thermal growth can result in microstructural change, which decreases long-term dimensional stability. It occurs particularly in certain aluminum alloys at elevated temperatures. T7 heat treatment is generally carried out to restrict thermal distortion (growth) of the alloy during operation. T7 heat treatment (overaged) provides a more dimensionally stable microstructure than T4 (naturally aged) or T6 (peak aged), and can reduce microstructural changes.[11] The strength changes with the grain size of castings, and generally, the thinner the casting, the higher the strength. The intermediate temperature strength of AC4B is sufficient, while the corrosion resistance is a little low due to the included Cu.

7.6   Conclusions

Valve springs use the superior characteristics of steel. It is possible to use Ti alloys to reduce weight, but steels will continue to be used for the majority of springs for the foreseeable future. The total balance of the system is crucial for the valve train. It is difficult to collect experimental data on the valve system during firing. Motoring testing is used to collect data instead, by turning an engine with an electric motor through the drive shaft. Measurements are used to optimize design. Comprehensive quality control is very important for all aspects of valve spring manufacture.

7.7   References and notes

1. Spring steels occur in two types. (i) The spring property results from heat treatment after shaping. (ii) The spring is shaped from pre-heat-treated steel. The latter occurs as piano wires, for which cold working gives the spring property, and oil-tempered wires, for which the spring property results from quenching and tempering. Piano wire has a microstructure of strained pearlite, while the oil-tempered wire has one of tempered martensite. Piano wire is likely to remain difficult to curl and hard to produce with a sufficiently thick diameter.

2. Chuo Spring Co., Ltd., Corporate Catalogue, (2003) (in Japanese).
3. Ibaraki N. R&D Kobe steel engineering report, 50(2000)27 (in Japanese).
4. Chuo Spring Co., Ltd., Corporate Catalogue, (1997) (in Japanese).
5. Takamura N., in Japan Society for Spring Research homepage, http://wwwsoc.nii.ac.jp, (2003) (in Japanese).
6. The spring limit value is defined as a limit stress after repeated deflection. Springs do not experience plastic deformation if used within the prescribed value. Copper-alloy springs use low-temperature annealing to raise the spring limit value. H. Yamagata and O. Izumi: *Nippon Kinzoku Gakkaishi*, 44 (1990) 982 (in Japanese). Cold working (including secondary working such as drawing or bending) is likely to cause stress corrosion cracking due to high residual stress. The season cracking in brass is well known. Low-temperature annealing is effective as a countermeasure.
7. Shot peening can introduce higher residual stress, as the original hardness of the worked piece is higher. It also prevents heat checking of casting molds (hot die steel). Shot peening technology resulted from research by GM.
8. Suto H., *Zanryuouryokuto Yugami,* Tokyo, Uchida Roukakuho Publishing, (1988), 98 (in Japanese).
9. *Metals Handbook* 8th ed, vol. 1, Ohio, ASM, (1961) 163.
10. The applied stress changes the spacing of crystal lattice planes. X-ray diffraction techniques can count the direction and quantity of the principal stress through measuring changes in spacing.
11. Boileau J.M., *et al.*, SAE Paper 2003-01-0822.

The crankshaft

8.1 Functions

The crankshaft converts reciprocative motion to rotational motion. It contains counter weights to smoothen the engine revolutions. There are two types of crankshaft, the monolithic type (Fig. 8.1), used for multi-cylinder engines, and the assembled type (Fig. 8.2) fabricated from separate elements, which is mainly used for motorcycles. The type of crankshaft determines what kind of connecting rods are used, and the possible combinations of crankshafts and connecting rods and their applications are listed in Table 8.1.

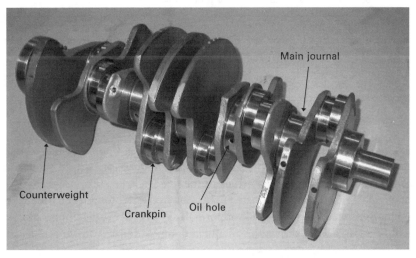

8.1 Monolithic crankshaft for a four–stroke engine. The fueling holes are for lubrication.

Crankshafts are made from forged steel or cast iron. Crankshafts for high-volume, low-load production vehicles are generally constructed from nodular

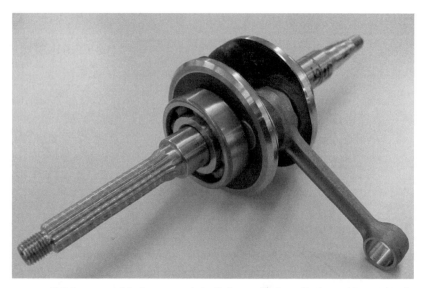

8.2 An assembly type crankshaft for a single-cylinder motorcycle. A connecting rod, a needle bearing and crankshaft bearings are already assembled.

Table 8.1 Combination of crankshafts with connecting rods. The monolithic crankshaft uses the assembled connecting rod, while the assembled crankshaft uses the monolithic connecting rod

Crankshaft type	Con-rod type	Engine
Monolithic	Assembly	Multi-cylinder four-stroke car engine, outboard marine engines
Assembly	Monolithic	Single- or twin-cylinder four-stroke engine, two-stroke engine

cast iron, which has high strength (see Appendix D). Fuel-efficient engines require a high power-to-displacement ratio, which has increased the use of forged crankshafts. The proportions of the materials used for crankshafts in car engines in 2003 were estimated to be, cast iron 25%, toughened (quenched and high-temperature tempered) or normalized steel 20%, and micro-alloyed steel 55%. Table 8.2 shows the chemical compositions of steel crankshafts.

8.2   Types of crankshaft

8.2.1   The monolithic crankshaft

Figure 8.1 shows a forged crankshaft for a four-stroke engine. The counterweight attached to the shaft balances the weight of the connecting

Table 8.2 Chemical compositions of crankshaft materials(%). JIS-S45C, S50C and S55C are plain carbon steel. In general, these are used in normalized state. JIS-SCM415, 420 and 435 are Cr-Mo steel, which are usually used in a quench-hardened state. The inside portion of a thick rod is unlikely to harden with quenching because of the slow cooling rate. Steels containing increased Cr and Mo can harden the deep inside portion of a thick rod

Chemical compositions	C	Si	Mn	P, S	Cr	Mo	V
JIS-S45C	0.45	0.25	0.8	0.03	–	–	–
JIS-S50C	0.5	0.25	0.8	0.03	–	–	–
JIS-S55C	0.55	0.25	0.8	0.03	–	–	–
JIS-SCM415	0.15	0.25	0.8	0.03	–	–	–
JIS-SCM420	0.2	0.25	0.8	0.03	1	0.2	–
JIS-SCM435	0.35	0.25	0.8	0.03	1	0.2	–
Micro-alloyed steel	0.5	0.25	0.8	0.03	–	–	0.1

rod (con-rod) and piston, to smooth revolutions. The con-rod rides on the crankpin via a plain bearing. The main bearing of the crankcase supports the main journal of the crankshaft.

The deep grooves in monolithic crankshafts are obtained by hot forging (Table 8.1). Carbon steels such as JIS-S45C, S50C or S55C with normalizing or toughening are used. Cr-Mo steel (typically, JIS-SCM435) and Mn steel are used to increase the strength. An alternative method using micro-alloyed steel containing V is becoming more common, as it is cheaper and does not require additional quench-hardening.

The intricate shape of the crankshaft requires a great deal of machining. It is common for about 0.1% lead or sulfur to be added to the base steel to improve machinability,[1] to make what is known as free-cutting steel. Figure 8.3 shows the microstructure of S50C-based leaded free-cutting steel after normalized heat treatment. Figure 8.4 is a sulfured steel with annealing. Included lead or MnS particles significantly function as a chip breaker and a solid lubricant and increase machinability.

Mass-produced sulfured steel is the oldest free-cutting steel. The sulfur is distributed homogeneously in the steel as MnS inclusions, which elongate according to the direction of rolling. As a consequence, elongation and impact strength in the direction transverse to rolling are weak. The machinability of this steel is proportional to the amount of sulfur it contains. Steel for high strength applications needs to contain less than 0.12% sulfur. Leaded free-cutting steel has isotropic properties in comparison with sulfured steel and is used for parts requiring high strength. The disadvantage of this steel is low fatigue life under rolling contact conditions. Crankshafts are normalized or quench-tempered after machining. To increase fatigue strength, induction hardening, nitrocarburizing and deep rolling are frequently employed.

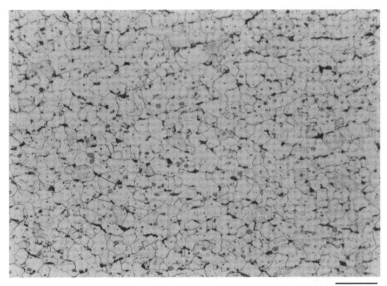

100 μm

8.3 Normalized microstructure of S50C leaded free-cutting steel containing 0.2% lead. Globular lead particles of a few μm disperse, while the matrix microstructure is not so clear due to weak etching.

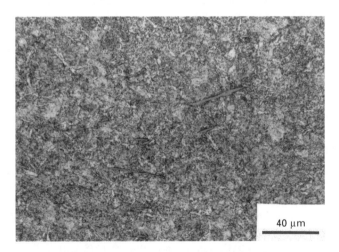

40 μm

8.4 Normalized microstructure of S50C sulfured free-cutting steel containing 0.06% sulfur. The MnS is elongated like thin sheets in the pearlite matrix. Chips break at the position of MnS or lead during machining, so that the chip does not tangle around the cutting tool.

8.2.2   The assembled crankshaft

Figure 8.2 shows an assembled crankshaft from a motorcycle, including the connecting rod and crankpin. The crankpin is precisely ground and force-fitted into the crankshaft body. The disassembled state is shown in Fig. 8.5. The appropriate fitting allowance and surface roughness give sufficient torque. To raise the torque, knurling, induction hardening or carburizing is often carried out around the hole.

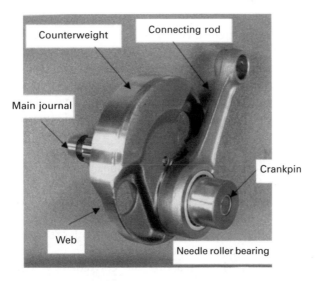

8.5 Disassembled crankshaft with the other web removed to show the big end.

This type of crankshaft is used in single or twin-cylinder engines for motorcycles. In two-stroke engines,[2] the structure has less lubrication oil at the crankpin bearing, and so it uses needle roller bearings. In low-output engines, the crankshaft body, including the shaft portion, is made from toughened plain carbon steel, such as JIS-S45 C or S55 C. The toughening process consists of quenching and high temperature tempering (see Appendix F). Additional induction hardening (described below) partly hardens the shaft portion.

Needle roller bearings (see Chapter 9) run on the surface of the crankpin. The high Hertzian stress caused by the rolling contact leads to fatigue failure at the pin surface. Therefore, a carburized Cr-Mo steel JIS-SCM415 or SCM420 (described below) is used. A bearing steel with a higher carbon content may also be used (SUJ2; see Chapter 9).

8.3   Rigidity

Monolithic crankshafts appear to have a high rigidity. However, the crankshaft is simultaneously subjected to bending and torsion when revolving. Under these conditions, it tends to wriggle like an eel,[3] and failure can occur as a result of fatigue. The main bearing clearance can be as small as 70 μm, but under these circumstances, the crankshaft deflects fully within the clearance while revolving. The trend towards reducing crankshaft weight means that the main bearing portion supporting the crankshaft is less rigid. This weakened main bearing cannot support the crankshaft sufficiently, which creates a severe fatigue situation.

The crankshaft is subjected to two types of stress, static and dynamic. Combustion pressure, inertial forces of the piston and con-rod, bearing load and drive torque all cause static stress. The vibration causes dynamic stress. If it occurs at the resonating frequency, the deformation will be very high and will instantly rupture the crankshaft. In order to achieve good acceleration, the crankshaft must have high static and high dynamic rigidity as well as low weight.

Modern engines are designed with size and weight reduction in mind. A short and small crankshaft makes the engine compact and then allows other components such as bearings and pins to be designed and built smaller, providing an overall reduction in system weight and associated cost savings.

While a cast iron crankshaft is less expensive, the lower rigidity of cast iron may allow abnormal vibrations to occur, in particular resonance, which is likely to appear at lower rotational velocities when the rigidity of the crankshaft is low. At the design stage, this can be avoided by increasing the crankpin diameter. However, raising rigidity in this way increases weight. Alternatively, an increase in rigidity of more than 10% can normally be gained by using steel instead of cast iron. Steel crankshafts have better potential to reduce noise levels and harshness over the entire engine revolution range, and careful design can make their use possible.

8.4   Forging

8.4.1   Deformation stress

The intricate shape of the crankshaft can be formed through hot forging using steel dies. In a red-heat state, steel behaves like a starch syrup and is extremely soft, so it molds easily to the shape of the forging die. Figure 8.6[4] compares the deformation stress of a steel at two strain rates. The stress required for deformation is low at high temperatures and hot forging takes advantage of this soft state. By contrast, deformation at low temperatures requires high stress, and the applied strain makes steel hard (known as work hardening, see Fig. 8.7). Deformation increases the dislocation density in the

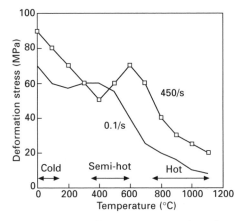

8.6 Influence of temperature and strain rate on the strength of carbon steel S35C. Dynamic strain ageing causes the peak around the intermediate temperatures from 400 to 700 °C The characteristic temperature range used for each forging process (cold, semi-hot or hot) is indicated. Steels recrystallize above 700 °C. The forging above this temperature is referred to as hot forging. At elevated temperatures, the deformation speed significantly influences the deformation stress. In general, the higher the speed (strain rate), the more the curve shape shifts to the higher temperature range. The normal forging machine gives stroke speeds of 0.1 to 1/s by the strain rate value.

steel (see Appendix G), which causes hardening. The crystal grains of steel have equiaxed shapes after the annealing and prior to deformation, but they stretch heavily after deformation (Fig. 8.7). Forging at low temperature (cold forging) cannot shape the deep grooves necessary for crankshafts and the die cannot withstand the load because work hardening dramatically increases the required load.

8.4.2   Recrystallization and recovery

Metals strained at low temperature undergo changes when heated. Figure 8.8 illustrates the hardness changes caused by heating. Hardness does not change when the temperature is low, but rapidly decreases above temperature T1. Changes in hardness are accompanied by microstructural change caused by recrystallization.

Heavy deformation at low temperature leaves the metal hardened and the microstructure changed, as shown in Fig. 8.7. Recrystallization creates new crystal grains in the strained matrix, which eliminates strain in the microstructures and causes softening (Fig. 8.8). The hexagonal pattern (grain boundary) indicates that the metal has recrystallized and that new crystals have been generated. Recrystallization substantially decreases dislocation

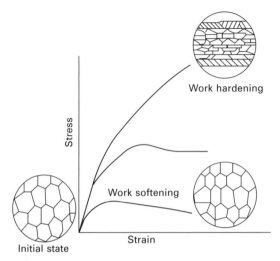

8.7 Temperature dependence of the stress-strain curve. The three curves correspond to the high, intermediate and low deformation temperature from the bottom. The illustrations indicate crystal grain shapes. An annealed microstructure containing equiaxed grains is on the left circle. Deformation changes it to the grain shapes shown in the right circles. At high temperature, dynamic recovery and dynamic recrystallization take place, which soften steel. The microstructure after deformation shows equiaxed grains when recrystallization takes place. By contrast, the large deformation at low temperature makes grains elongated shapes. Metal hardens with increasing strain and softening does not take place. The hardening is called work hardening or strain hardening.

density. Each metal has a specific minimum temperature (T1) at which recrystallization takes place.

When recrystallized metal is annealed further at a higher temperature above T2 (Fig. 8.8), the recrystallized grains grow. Below T1, recrystallization does not take place, and a rearrangement of dislocations along with a decrease in density occurs, resulting in slight softening. This is referred to as recovery. Plastic working carried out above the recrystallization temperature T1 is generally called hot working. The temperature at which recrystallization occurs is different for each metal, the recrystallization temperature of steel is around 700 °C.

Hot forging of steel is carried out at the red heat state, above 700 °C (Fig. 8.6). During hot forging, steel goes through recrystallization and recovery as well as strains. These softening processes remove the accumulated strain and thus the steel does not harden (Fig. 8.7), making shaping easy. The recrystallization and recovery that take place during hot working are referred to as dynamic recrystallization[5] and dynamic recovery, respectively. These processes eliminate work hardening despite the heavy deformation produced

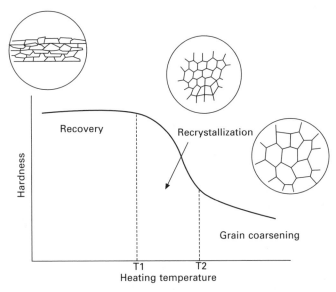

8.8 Deformed metal softens with heating. The grains after deformation (at the top left) are fully extended. The heating of the deformed metal above the recrystallization temperature (T1) causes recrystallization, which removes the deformed microstructure and generates recrystallized grains (at the center). Below T1, microstructure does not change apparently, but dislocations in the metal rearrange. The heating (annealing) at higher temperatures (above T2) grows the recrystallized grains (on the right). The grain growth (coarsening) decreases hardness. The smaller the resultant grain size, the higher the hardness (strength). Hence, overheating during annealing should be avoided.

by forging, giving malleability. Recovery and recrystallization that take place in cold-worked metal during annealing are different, and are called static recovery and static recrystallization, respectively.

8.4.3   Hot forging

Figure 8.9 illustrates the die forging process for a monolithic crankshaft.[6] The steel bar is first sheared into a billet to adjust the weight. Induction or gas heating heats the billet to around 1,000 °C, using rapid induction heating, which causes less decarburization or oxidation.

Rough forging distributes the material thickness along the axis. Shaping by a forging roll and bending are then carried out simultaneously. Die forging then forms the intricate shape, and finish forging adjusts dimensional accuracy. Burr shearing removes the flash from the shaped material, and the shaped material is then straightened to remove the bend. These processes are carried out at redheat and the shaped material is machined after cooling.

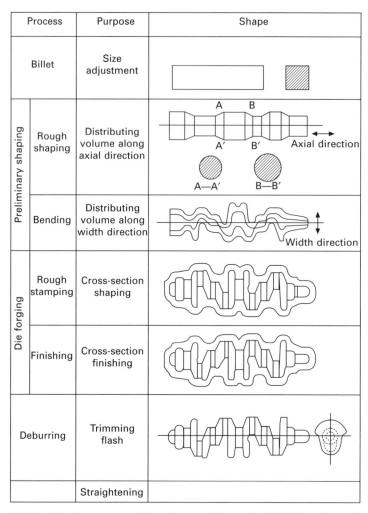

Process		Purpose	Shape
Billet		Size adjustment	
Preliminary shaping	Rough shaping	Distributing volume along axial direction	
	Bending	Distributing volume along width direction	
Die forging	Rough stamping	Cross-section shaping	
	Finishing	Cross-section finishing	
Deburring		Trimming flash	
		Straightening	

8.9 Hot forging process for a four-stroke crankshaft. The sheared billet experiences six processes from initial rough forging to final straightening. In the rough shaping stage, the forging roll shapes the billet into the stepped shaft. The fiber flow (described later) is schematically illustrated in the figure of bending process. The shaped material is finally straightened if it is distorted.

A water-soluble lubricant is splashed on the die surface to cool it. The rapid heating and cooling of the die shorten its life, to around 10,000 shots or less. Hot forging can shape intricate forms, but cannot give high dimensional accuracy because oxide scale accumulates on the surface, so the shaped material must be milled to give the required shape.

Crankshafts need high-strength materials, but these generally have low forgeability or machinability, and as a result are costly to use. Cr-Mo steel gives higher strength, but the higher deformation resistance shortens die life. Figure 8.10 shows the process design sheet listing the key factors in forging.[7] The following keep forging costs low:

1. Forging at low temperature without heating.
2. Using soft materials that reduce forging loads, resulting in a smaller machine and longer die life.
3. Shallow shapes that require shallow grooves carved in the die do not generate high thermal stress, which lengthen die life.
4. Ensuring low ratio of product to flash (low mold yield), where forging defects such as material lapping are less likely to occur.
5. Using very high production numbers for a small number of items, which means die changes are fewer and therefore operational downtime is reduced.

Despite the requirements of manufacturers to keep costs low, it is always the market that determines material, shape and production numbers. A skilled forger can optimize the process by considering several of the factors listed above and in Fig. 8.10.

The forging machine specifications are determined by the necessary dimensional accuracy and the quantity of products. There are different types of forging machine classified according to the drive system, for instance, air hammer, mechanical press, hydraulic press, etc. Crankshafts are forged mainly using air hammers and mechanical presses. Hot forging with an air hammer[8] requires a shorter time for die changes. The die and machine are less costly, but require a skilled operator as the material is handled manually. This is appropriate where production numbers are small.

In the manufacturing process of assembled-type crankshafts, upset forging first swells the web and counterweight from a bar. One end is heated for upsetting, using the same process as that for valves (see Chapter 6). Following this, die forging gives the final shape. Various forging methods[6] are listed in Fig. 8.11.

8.4.4   Cold and semi-hot forging

Cold forging is carried out at ambient temperatures. Typical forging patterns are illustrated in Fig. 8.12 – backward extrusion, forward extrusion and upsetting. In shaping a cup from a disk-shaped billet (a), the forward extrusion (c) pushes out the crown in the travelling direction of the press punch, while the backward extrusion (b) pushes out the crown in the opposite direction. Upsetting (d) expands the billet.

Cold forging does not generate oxide scale because of the low temperature, so it produces accurate, near-net shapes without flash. The metal can be shaped completely within a closed die, and production numbers have increased

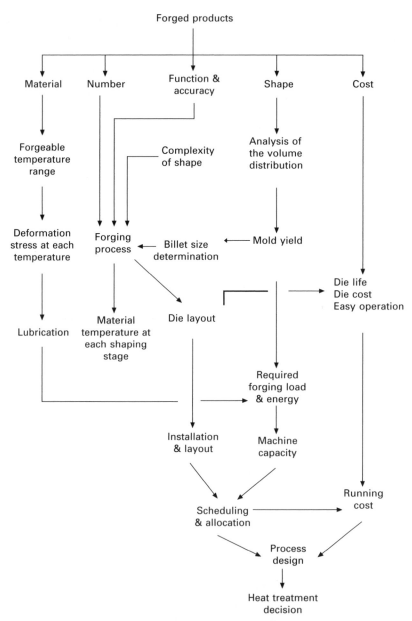

8.10 Design of forging process. The function of the forged part determines the required shape, material, accuracy and so on. For the mechanical designer, the accuracy and strength of the part is important, while the forger has various restrictions. These are that the material is special to the market or that the die cannot withstand severe shaping. Hence, the final shape results from a compromise. An excellent part is made through the collaboration of the engineers who know mutual needs well.

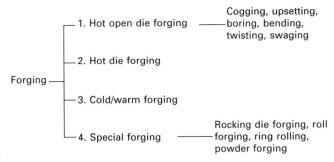

8.11 Classification of forging processes. The open die forging uses simple tools instead of dies.

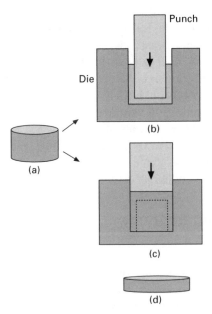

8.12 Cold forging illustrating (a) initial billet, (b) backward extrusion, (c) forward extrusion and (d) upsetting.

significantly in recent years. This method reduces costs because the near-net shaping using a closed die decreases the need for additional milling, but the high forging force necessary can shorten die life because of the high stress. Cold forging is therefore applied mainly to simple shapes such as shafts or round cups. Asymmetric or complicated shapes, such as crankshafts, are still produced by hot forging.

Phosphate conversion coating is frequently applied as a solid lubricant on the billet surface during cold forging. The decrease in friction reduces the forging force. To increase the malleability of steel, spheroidizing annealing is often carried out (see Appendix F). This annealing modifies the lamellar carbide to a round shape, which prevents micro-stress concentration and thus avoids rupture of the workpiece even under severe straining.

Semi-hot or warm forging is carried out at intermediate temperatures of around 300–600 °C (Fig. 8.6). The semi-hot state of the billet decreases deformation stress. This process is similar to cold forging in that low levels of oxide scale keep the accuracy high. It is applied to high strength materials or large parts.

8.4.5   Combination forging

Closed die forging can produce near-net shapes with minimal flashing. However, a high forging force is needed to form the shape. Combination forging uses rough shaping by hot forging followed by cold forging. If a part can be produced by several different methods, then production volume determines the most suitable forging process. Precision forging using a closed die can produce a near-net shape. Although the precision die is expensive, less milling is required and so precision forging is less costly for large production volumes. For small production volumes, rough hot forging and finishing by milling are generally used because the die for hot forging is cheaper. Ring rolling, which shapes annular rings through rolling, and powder forging, which raises the density of sintered parts, are among the special forging methods required in some situations (Fig. 8.11).

8.5   Surface-hardening methods

8.5.1   Carburizing

Steel automobile parts operate under sliding or rolling conditions and are highly stressed at the surface. Various surface hardening processes are used, and carburizing is a typical case-hardening process used for pins and gears. The single or twin-cylinder engine generally uses a needle roller bearing at the big end of the connecting rod (see Chapter 9). The big end works as an outer raceway for the rollers and the crankpin as an inner raceway. The rolling contact that these surfaces are subjected to results in a high Hertzian stress. Both big end surfaces of the connecting rod and crankpin are generally carburized to raise surface hardness to counteract this stress. Carburizing gives very high hardness and is resistant to wear, but surface failure can occur if operating conditions are too severe.

Figure 8.13 shows pitting failure at the surface of a crankpin. Figure 8.14 shows the microstructure underneath the pitting, 50 μm below the surface. Pitting (Chapter 5) is a typical fatigue phenomenon that occurs under high contact pressure. The white microstructure contains cracks, and these cracks

5 mm

8.13 Pitting observed at a crankpin surface.

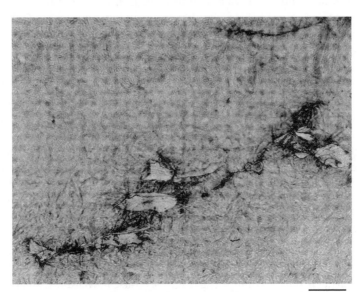

25 μm

8.14 Magnified view of a fatigue crack at a position 50 μm below the surface. This type of crack often accompanies hard white regions which initiate pitting. The white portion called the white etching region consists of hard ferrite and pearlite.

are the cause of the pitting. This type of failure is likely to appear in shallow portions, where Hertzian stress is at its maximum. These cracks often accompany white or dark regions under microscopy. These are clearly observed after the chemical etching of the microstructures and called white etching or dark etching regions. A butterfly shaped microstructure is also observed. These microstructures, characteristic of surface failure, are often generated at the periphery of nonmetallic inclusions. Recent research[9] has revealed that the white etching region consists of ferrite or pearlite. Local heating by stress concentration raises the temperature of the region to as high as 600 °C, which generates the white etching region.

Figure 8.15(a) shows a carburized microstructure of a Cr-Mo steel, JIS-SCM420. Carburizing is carried out by exposing a part made of low carbon steel for a defined period of time in a hot atmosphere with a high carbon concentration (CO), which enriches the carbon concentration at the surface (Fig. 8.16). Then the part is quenched in water or oil, whereupon a martensitic transformation takes place at the surface because of the rapid cooling and the high carbon concentration on the surface. The part does not undergo a martensitic transformation internally because of the slow cooling rate and low carbon concentration away from the surface, thus hardening takes place only at the surface and the center of the part stays soft.

Figure 8.17 shows a typical hardness distribution curve for a carburized part according to depth from the surface, showing a hardness of 710 HV at 50 μm below the surface. Two values representing the hardness distribution after carburizing are generally used for quality control, effective case depth and total case depth. The effective case depth is the thickness of the portion showing a higher hardness than the prescribed value.[10] The total case depth is the thickness showing a higher hardness than the base steel. These values measured in the hardness distribution curve guarantee the carburizing heat treatment.

Carburizing consists of a series of heat treatments; carburizing, quenching and tempering. Figure 8.18 explains the principle of carburizing using the iron-carbon phase diagram. The actual procedure is illustrated in Figure 8.19. First, the carbon concentration is increased to around the eutectoid point (0.8% C) (represented by the solid arrows in Fig. 8.18), in the process known as eutectoid carburizing. The atmosphere in the furnace supplies the carbon atoms, and when it is sufficiently high, the carbon atoms spontaneously permeate into the steel. Carbon atoms rapidly diffuse at temperatures above A_1. Then quenching and tempering (Fig. 8.19) raise the hardness and toughness of the carburized steel through martensitic transformation.

The austenite phase can dissolve a larger amount of carbon. Excessive carburizing beyond the eutectoid point (represented by the broken arrows in Fig. 8.18) generates carbides and is called supercarburizing (described below).

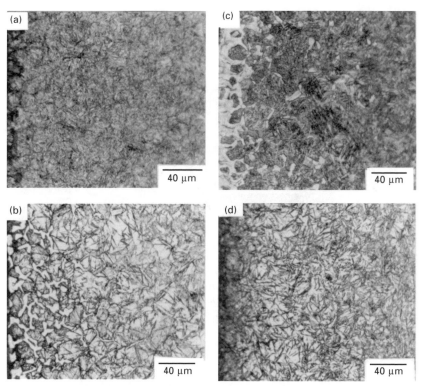

8.15 Microstructure of carburized SCM420. (a) Quenched microstructure after eutectoid-carburizing. The black portion near the surface is troostite consisting of fine ferrite and carbide. The portion shows an unwanted microstructure generated by imperfect quenching. Grain boundary oxidation locally decreased alloyed elements, having obstructed quench-hardening. (b) Super-carburized quench-hardened microstructure. A netlike carbide is observable at 60 μm depth from the surface, being the same microstructure as a hyper-eutectoid steel. (c) Decarburized microstructure in a carburized layer. (d) A carburized microstructure containing retained austenite. Generally, the microstructures (b), (c) and (d) are defective. Unlike martensite, the carbide in (b) does not dismantle even at high-temperature annealing. Hence, it can give wear resistance at high temperature. Also, the retained austenite (d) increases fatigue strength.

Compressive residual stress generated by carburizing

Martensitic transformation is accompanied by lattice expansion. This produces a favorable compressive residual stress at the surface and significantly increases fatigue strength. Figure 8.20(a) shows the mechanism. Martensitic transformation starts at the Ms point (°C) upon cooling. The higher the carbon concentration of a steel, the lower the Ms point of the steel.[11] The empirical equation showing the relationship between Ms temperature and alloying elements is

$$\text{Ms (°C)} = 550 - 361 \times (\text{C\%}) - 39 \times (\text{Mn \%}) - 35 \times (\text{V \%}) - 20 \times (\text{Cr \%})$$

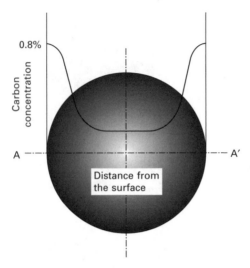

8.16 Distribution of carbon concentration in a carburized shaft. The carbon concentration is indicated at A-A' of the shaft cross-section. A high carbon concentration is observable only at the surface area. The central low carbon area corresponds to the base steel.

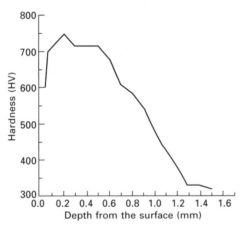

8.17 Hardness distribution of carburized Cr–Mo steel SCM420. The amount of retained austenite which has not transformed to martensite measures 20%. The soft surface layer of about 100 μm is polished off in finishing.

$- 17 \times (\text{Ni \%}) - 10 \times (\text{Cu \%}) - 5 \times (\text{Mo\%} + \text{W\%}) + 15 \times (\text{Co \%}) + 30 \times (\text{Al \%})$.

In Fig. 8.20(a), the two cooling curves illustrated correspond to the slow cooling rate inside a part and the higher cooling rate at the surface. The carbon-enriched surface has a lower Ms point (Ms at S) in comparison with the internal portion, which has a lower percentage of carbon (Ms at I).

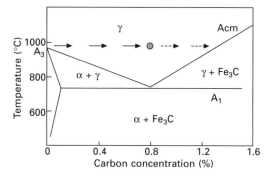

8.18 Principle of carburizing. Carburizing increases carbon concentration towards the arrow direction. If the carbon density exceeds the eutectoid point of 0.8% C, unwanted cementite appears after cooling. Carburizing should terminate within 0.8% C.

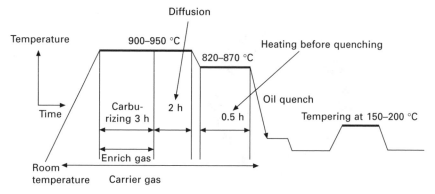

8.19 Carburizing treatment diagram. Carburizing takes place at temperatures from 900 to 950 °C. The part just after carburizing is too hard and brittle. The additional tempering at 150 to 200 °C increases toughness.

Martensitic transformation expands the iron lattice. In the quenching stage, internal cooling is slower than surface cooling (Fig. 8.20(a)). Hence, martensitic transformation takes place first internally, at time tI (< tS), or it does not appear at all if the cooling rate is too slow. Martensitic transformation then takes place at the surface at tS. Expansion of the surface as it transforms is restricted by the expansion that has already taken place internally, and this restricted expansion generates a compressive residual stress in the hard surface layer (Fig. 8.20(b)).

If a part has both transformed and untransformed areas, the resultant difference causes unacceptable distortion. If the relaxation cannot accommodate the distortion, stress is generated, and a large stress may break the part during quenching. This is known as a quenching crack.

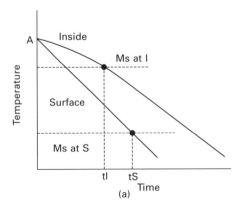

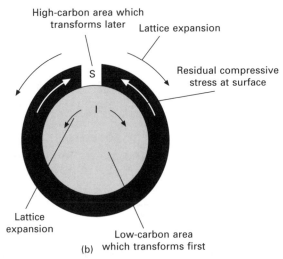

8.20 Occurrence of compressive residual stress by martensitic transformation. (a) Cooling curves of the inner portion I and of the surface portion S. The surface portion cools rapidly. (b) Compressive stress in the surface. The inner portion I has a lower carbon concentration compared to the carburized surface S. The inner portion having a low carbon concentration first transforms at time tI (Ms at I). The surface portion transforms at later time tS (Ms at S). The pre-transformed inner portion restricts the later surface expansion, which generates a residual compressive stress of the surface.

Gas carburizing

Gas carburizing is carried out in a gaseous atmosphere containing CO and CH_4. The gas dissociates catalytically at the hot steel surface to generate elemental carbon atoms. This carbon permeates into the steel lattice to form a carbon-enriched surface layer (Fig. 8.16).

Producing the gas for carburizing is typically conducted in stages. First, a propane or natural gas is mixed with a calculated quantity of air. The mixture burns incompletely when passed over a hot Ni catalyst at about 1050 °C. The resultant gas is cooled and dehydrated to become a mixed gas (endothermic gas), with a typical ratio of CO: H_2: N = 23:31:46. To raise the carbon concentration (carbon potential) further on the surface, a little hydrocarbon is added to enrich the gas. It is relatively easy to control the temperature and atmosphere during gas carburizing, and this process is suitable for mass production.[12]

A drip-feed furnace uses an alternative method to generate the gas. A small amount of alcoholic gas, such as CH_3OH, is fed directly into the furnace and dissociates to generate the carburizing gas. Although a gas generating system is not necessary, the gas composition in the furnace must be carefully controlled. This simple method is suitable for medium production volumes.

The entire surface of the part can be carburized by this process, but some areas may not need carburizing, and these can be coated to prevent carburizing taking place. For example, the assembled crankshaft (Fig. 8.5) needs carburizing only at the shaft portion. The portion without the shaft is copper-plated to prevent carbon penetration during carburizing.

If the depth of case hardening is too shallow, delayed failure can occur. This is the phenomenon where a fracture suddenly occurs after heat treatment is terminated. It happens particularly in high-strength parts such as bolts. Use of highly alloyed steel raises the effective case depth and prevents delayed failure.

The carburized steel part is resistant to the repetitive small stresses that result in fatigue and wear. However, the hard surface is brittle and weak against large impact loading.[13] Carburizing is sometimes carried out on an entire crankshaft to raise fatigue strength, but care must be taken to avoid straightening of the distortion caused by carburizing treatment.

Surface hardening treatment is called case-hardening, while the hardening of the entire part is through-hardening. Steel suitable for case-hardening is referred to as case-hardening steel.

Vacuum carburizing

The low-pressure process of vacuum carburizing[14,15] is becoming more widely used. This treatment uses hydrocarbon gases such as acetylene (C_3H_8), ethylene (C_2H_4) and/or propane (C_2H_2), and is carried out in the low-pressure range between 2 and 50 kPa. There is no CO_2 emission.

The part is initially heated in low-pressure nitrogen (around 150 kPa), followed by heating in a vacuum to a carburizing temperature of between 900 °C and 1050 °C. The carburizing gas is then admitted through jets and thermally dissociates to generate elemental carbon. The hydrogen by-product reduces metal oxides on the surface, which facilitates absorption of carbon into the steel.

The carburizing process itself comprises carburizing and diffusion. In the first stage, the inflow of carburizing gas provides a very high concentration of carbon that can be absorbed by austenite. In the diffusion stage, the gas inflow is cut off, and the carbon is allowed to diffuse into the surface. Diffusion reduces the surface carbon concentration, which allows for further carburization. Surface oxidation is avoided because oxygen-free gases are used.

The carburizing and diffusion stages are timed and the process can be continuous (single pulse) or have repeated carburizing and diffusion steps (a multiple process). A high gas inflow and higher process temperature can shorten the carburizing process. The gas supply is modulated in accordance with the surface area of the part, which helps to ensure that there are no uncarburized or over-carburized areas.

High-pressure gas quenching is generally combined with vacuum carburizing. Quenching gas such as nitrogen or helium at 2 MPa is used. The quenching intensity is controlled by gas pressure and the parts are dry after quenching. This process does not require washing afterwards and therefore problems associated with waste water disposal are avoided.

In conventional quenching using liquid media, film boiling, bubble boiling and convection take place. The inhomogeneous quenching caused by these phenomena is likely to distort the part. However, gases show no phase changes during quenching and the homogeneous cooling helps reduce distortion. Vacuum carburizing with high-pressure gas quenching is a relatively new process, developed in the 1960s, and used increasingly to meet environmental requirements.

Abnormal microstructures occurring in carburizing

It is common for most steel parts to be heat treated. Contemporary heat-treating facilities are computer controlled and have excellent sensors, so there are relatively few rejects. However, abnormal microstructures in carburizing do still occur occasionally and these are discussed below.

Figure 8.15(b) shows a super-carburized microstructure with a hardness of 700 HV. White carbide is observable particularly at the grain boundaries, and has been generated by excess carburizing above the eutectoid point, 0.8% C (Fig. 8.18). The gray needle-like portion is martensite, whereas the white matrix is retained austenite. The amount of retained austenite at 100 μm below the surface is about 50%. The black portion near the surface is troostite, with a hardness as low as 578 HV.

This type of microstructure appears when the carbon potential is kept at temperatures above the Acm line for a long time (Fig. 8.18).[16] Troostite appears when the alloying elements are absorbed into the carbide and the grain boundary is oxidized. Once this carbide appears, even additional heat treatment cannot remove it.

The carburizing gas inevitably includes oxygen in the form of H_2O, CO_2 and CO, causing oxidation of grain boundaries. Oxygen diffuses rapidly along the grain boundary. The alloyed Si, Mn and Cr are likely to be oxidized at the grain boundary near the surface (Fig. 8.15(a)). This is inevitable in standard gas carburizing, but the defective layer can be removed by additional polishing. Grain boundary oxidation itself does not lower fatigue strength, but it does cause local decreases in the concentration of Si, Mn and Cr, which are important for increasing hardness. As a result, pearlite and/or bainite appear near grain boundaries when cooling is slow, and lower fatigue strength substantially.

Figure 8.15(c) shows a decarburized layer. The white portion in the surface layer, 0.1 mm deep, is ferrite formed by decarburization. In Fig. 8.15(c), decarburization has occurred after carburizing, when the carbon potential was lowered during cooling to 780 °C, and this has lowered the carbon quantity to 0.44% at a depth of 500 μm. This can occur when the carbon potential during cooling is controlled by forced air flow.

Figure 8.15(d) shows a microstructure with as much as 41% retained austenite. The hardness is 582 HV at a depth of 50 μm. The black part near the surface is troostite caused by grain boundary oxidation. A large amount of retained austenite is unacceptable for a precision part. If the unstable austenite transforms to martensite during operation, the shape changes and loses its dimensional accuracy. However, it has been reported that if retained austenite is less than 30%, this can increase surface fatigue strength under rolling contact conditions.[17]

8.5.2   Nitriding

Nitriding was originally promoted by A. Fry in 1923.[18] It is a case-hardening treatment carried out in enriched nitrogen. When a steel part is placed in a hot, nitrogen-rich atmosphere containing NH_3, the NH_3 decomposes at the steel surface to catalytically generate elemental nitrogen, which diffuses into the material. The nitrogen expands the iron lattice and also forms hard compounds (the nitrides Fe_4N and Fe_3N) with iron atoms. The expanded lattice and finely dispersed nitrides immobilize dislocations and so harden the steel surface. Unlike carburizing, quenching is not necessary after nitriding.

This treatment is carried out in the temperature range where α iron exists in the iron-carbon phase diagram. Normally it is around 500–520 °C. To obtain very high levels of hardness, 40 to 100 hours of nitriding are needed. Nitridable steel reaches the necessary hardness by forming stable nitrides, which requires alloying elements such as Al, Cr, Mo, V and/or Ti. Al gives high hardness, Cr increases the thickness of the nitrided layer and Mo suppresses temper embrittlement (even if the part is heated for a long time during nitriding). Productivity is low, so that this treatment is used only for special purposes at present. Conversely, nitrocarburizing is widely used for mass-produced parts.

8.5.3   Nitrocarburizing

Nitrocarburizing is another case-hardening process, and is also known as ferritic-nitrocarburizing, or cyaniding.[19] It is a modified nitriding process in which a gas containing carbon is added to the ammonia atmosphere. Steels held at high temperatures in this gaseous atmosphere absorb carbon and nitrogen simultaneously, at a temperature below A_1(around 560 °C), in the ferrite region of the phase diagram. The shorter time period as well as the lower temperature gives a shallow case depth, typically about 0.1 mm. The amounts of nitrogen and carbon in the layer are adjustable within certain limits.

Gas nitrocarburizing is suitable for mass-produced parts.[20] N and C are diffused under an atmosphere of 50% NH_3 and 50% RX gas (a transformed gas of propane and butane). Heat treatment at around 560 °C results in a hard surface containing Fe_3N.[21] The hardness can be adjusted by changing the time of treatment, from 15 minutes up to 6 hours. It is normally implemented in a tunnel-type furnace, where parts enter at one side and exit on the opposite side, but a batch type furnace may also be used. Figure 8.21 shows the hardness distribution of a Cr-Mo steel, JIS-SCM435, that has undergone gas nitrocarburizing.

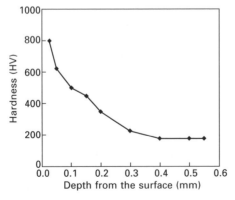

8.21 Section hardness distribution of gas nitrocarburized Cr-Mo steel SCM435. The quench-tempered sample is nitrocarburized for 3 h at 570 °C, followed by oil cooling.

Carburizing is implemented in the austenite region at around 900 °C and distortion[22] during heating and quenching is likely to occur. By contrast, nitrocarburizing is implemented at temperatures as low as 560 °C and does not cause martensitic transformation, distortion is, therefore, less after this treatment.

Liquid nitrocarburizing, also called cyaniding, is carried out in a molten salt bath, using a mixture of cyanides XCN, XCNO, and X_2CO_2 (X: Na or K). The hardening agents CO and elemental N are produced in the bath in the presence of air. It is possible, within limits, to regulate the relative amounts of carbon and nitrogen in the surface layer. The treatment time ranges from 15 minutes to 3 hours. This process gives a hard layer in alloys such as stainless steel, for which gas nitrocarburizing cannot give sufficient hardness. It is typically used for engine valves which have a high Cr content. Since it only requires a bath for molten salts, the facility is less costly, even when production numbers are small. However, its use is becoming less common due to the hazardous nature of the cyanide bath.

8.5.4   Carbonitriding

Figure 8.22 compares hardness against tempering time at 350 °C for different methods of case hardening. The carburized surface loses hardness when kept at temperatures above 200°C. In the figure, carburizing produces a greater decrease in hardness after two hours in comparison with the other methods. By contrast, case-hardened layers containing N (super carbonitriding and carbonitriding) lose hardness more slowly. This is due to the effect of the stable nitride compound dispersed in the matrix, whereas the martensite in the carburized layer rapidly loses hardness when heated above 200 °C.

Engine parts are sometimes exposed to high temperatures as well as high stress. For such a situation carbonitriding is very effective. This treatment

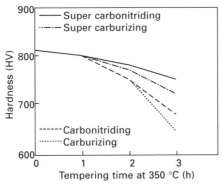

8.22 Hardness decreases of case hardend Cr-Mo steel SCM420 after tempering at 350 °C.

generates a carburized surface containing nitrides, giving a stronger surface at elevated temperatures than that obtained by normal carburizing. Carbon and nitrogen diffuse simultaneously in carbonitriding. Carbon enrichment is the main process, but nitrogen enrichment occurs if the nitrogen concentration in the gas is sufficiently high. The amounts of carbon and nitrogen in the layer are adjustable according to the composition of the gas and its temperature.

Carbonitriding has been found to be very effective at raising the strength of parts subjected to extremely high contact stress. This treatment is successfully applied in transmission gears[23] as well as ball and needle bearings. A carburized layer containing N has superior heat resistance as observed in Fig. 8.22, so it can withstand the heat caused by the high contact stress at the surface. Roller bearings of JIS-SUJ2 steel use carbonitriding at the austenite region to increase resistance to rolling contact fatigue (see Chapter 9).

Supercarbonitriding generates a carburized surface containing both nitrogen and globular carbide. It has been used successfully to give a long rolling contact fatigue life to the crankpin of an assembed crankshaft.[24]

8.5.5   Ion nitriding

Ion (plasma) nitriding makes use of an ionized gas that serves as a medium for both heating and nitriding. The parts are placed in a vacuum chamber and the furnace is filled with process gas containing N_2 and H_2 to a pressure of 100–800 Pa. The plasma is created through glow discharge by applying a direct electrical current, with the part acting as the cathode and the chamber wall acting as the anode. The applied voltage (300–800 V) accelerates the ions towards the surface of the part. The plasma process operates at temperatures between 400 and 800 °C and the treatment is generally implemented by batch. It is frequently used for forging dies or casting molds to raise resistance to wear and thermal fatigue.

Vacuum plasma carburizing has been investigated. This process is similar to the ion nitriding process. Plasma carburizing using methane is a special process for partial hardening and carburizing of internal bores. The plasma is created between the part as cathode and the chamber wall as anode. For partial carburizing, the plasma effect may be prevented by covering with metallic conducting masks or sheet metal where it is not required. The plasma cannot develop under the cover and therefore the covered surface remains free of carburizing.

Table 8.3 summarizes the major case-hardening processes. The terminology of carbonitriding and nitrocarburizing often creates misunderstandings. Carburizing is the term for adding only carbon. In carbonitriding, the main element is carbon with a small amount of nitrogen. The dopant in nitriding is nitrogen alone. In nitrocarburizing, the main dopant is nitrogen but a small amount of carbon is added simultaneously. For carburizing and nitriding, the

Table 8.3 The difference in dopants in case hardening

Case hardening	Dopant	Temperature
Carburizing	C	High temperature above A_1
Carbonitriding	C + N (small amount)	
Nitriding	N	Low temperature below A_1
Nitrocarburizing	N + C (small amount)	

difference is clear. On the other hand, carbonitriding and nitrocarburizing are frequently used with the same meaning. The terminology 'austenitic nitrocarburizing' is also used.

8.5.6   Induction hardening

The surface methods described above include thermal treatments with chemical changes. The following methods may be classified as simply thermal treatments without chemical change. They can be used to harden the entire surface or localized areas. Some methods heat only the surface of a part. If a part made of high carbon steel is heated to austenite only at the surface, the subsequent water quenching transforms the surface into martensite to raise hardness at the surface.

Flame-hardening consists of austenitizing the surface by heating with an oxyacetylene or oxyhydrogen torch and immediately quenching with water. This process only heats the surface so that the interior core does not change. This is a very convenient process and is sometimes used for surface-hardening large dies with air cooling, since highly alloyed tool steel for dies hardens even in air cooling. However, managing the hardness can be difficult.

Induction hardening is an extremely versatile method that can produce hardening over an entire surface, at a local surface or throughout the thickness. A high-frequency current generated by an induction coil heats and austenitizes the surface, and then the part is quenched in water. The depth of heating is related to the frequency of the alternating current; the higher the frequency, the thinner or more shallow the heating. Tempering at around 150 °C is subsequently carried out to increase toughness. Induction heating is also used for tempering after quenching. The monolithic crankshaft uses induction hardening[25] of the crankpin and the corner radii between the crankpin and web. The assembled crankshaft uses induction hardening of the hole into which the crankpin is forcefitted and of the corner radii.

Figure 8.23 shows the hardness distribution[26] of steel JIS-S50 C normal to the surface. Figure 8.24(a) schematically illustrates the microstructures generated by induction hardening. The induction coil is also shown. The hardened microstructure shows a pattern (quenching pattern) when the cross-section is chemically etched, as shown in Fig. 8.25. Martensitic transformation

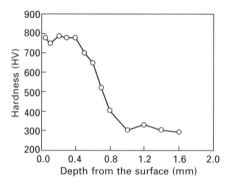

8.23 Cross-sectional hardness distribution of induction-hardened carbon-steel JIS-S50C.

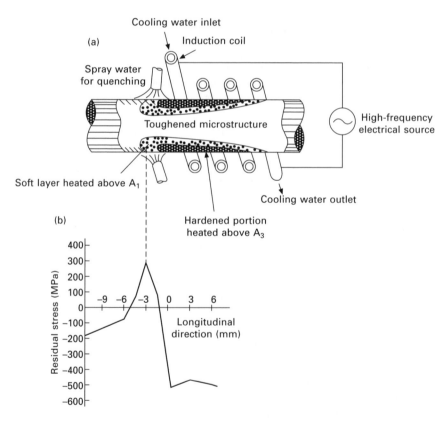

8.24 (a) Induction-hardened pattern in the cross cut view of a rod.
(b) The residual stress distribution along the longitudinal direction.

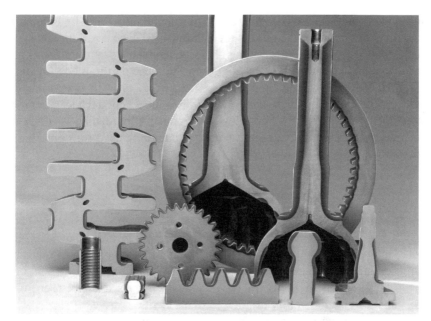

8.25 Cross-sectional view of induction hardened parts (courtesy of Fuji Electronics Industry Co., Ltd). The hardened portions at the surfaces are distinguishable by etched contrast due to the difference in microstructure. A crankshaft is shown on the left. The pin and the fillet between the web and pin are hardened.

expands the crystal lattice of the surface, whereas the untransformed internal portion restricts the expansion. This restraint leaves a compressive residual stress[27] in the surface and such stress raises wear resistance and fatigue strength.

Induction hardening gives high hardness at the surface, but is accompanied by an undesirable soft area just under the surface. The softened layer appears broadly at the boundary between the hardened area and unhardened area. It also appears at the surface (Fig. 8.24(a)) where the induction hardening is terminated. The soft area is caused by incomplete austenitizing near point A_1.

Generally, this layer has a tensile residual stress. Figure 8.24(b) shows a residual stress distribution in the longitudinal direction measured by X-ray. A high tensile stress is observable at the edge (– 3 mm from the point 0) of the hardened area, and a stress concentration at the edge is likely to initiate fatigue cracking. Therefore the edges of the hardened area must never be near fillets, notches or grooves, so these must be included in the hardened area. Weaknesses can be avoided by adjusting the shape of the part or the quenching pattern.

Induction hardening is a short-term heat treatment method, and it must be ensured that the initial microstructure can transform rapidly into homogeneous austenite during heating. Normalizing or toughening prior to induction hardening can decrease the dispersion of hardness at each position. Since the installation for induction hardening is compact, hardening can be implemented in the machining line without having to transport the part to a heat-treating plant. If the part is completed without tempering, or with tempering by induction heating, a build-up of stock waiting for the additional heat treatment is avoided and cost is lowered. However, induction hardening is likely to distort a thin and long crankshaft, and so it is mainly used for crankshafts with a thick crankpin diameter.

8.6   Micro-alloyed steel

The heat treatments described above can improve desirable properties, but they also raise costs. Recent cost-saving measures have included the increasing use of micro-alloyed high-strength steel instead of the conventional quench-hardened steel for crankshafts. Developments in manufacturing techniques and in alloyed steels have led to improved strength, increased fatigue properties and enhanced machinability in micro-alloyed steels.

Precipitation hardening is the main method for increasing strength at the cooling stage after hot forging. Micro-alloyed steel contains a small amount of vanadium (see Table 8.2), which dissolves in the matrix during hot forging above 1,200 °C. During air cooling, the dissolved V combines with carbon and nitrogen to precipitate as vanadium carbide and nitride at around 900 °C. Tempering after air cooling is not necessary because these precipitates in the ferrite and pearlite matrix strengthen the steel (see Appendix F). Maintaining the required temperature for a period of time after hot forging ensures sufficient precipitation. Typically, spontaneous cooling from 1,200 °C to 300 °C for a large crankshaft weighing 32 kg takes about one hour, and hardening occurs during this cooling period.

Figure 8.26 shows the relationship between cooling rate, hardness and tensile strength. Controlling both forging temperature and cooling rate adjusts hardness and strength to obtain the required values. For example, a 100 mm diameter rod has a cooling rate of 10 °C/min from 1,200 °C. The diagram indicates that hardness for this rod at this cooling rate will be around 280 HV, and the tensile strength around 900 MPa. In the range given in Fig. 8.26, the faster the cooling rate, the higher the hardness. This is because higher cooling rates give a finer pearlite matrix, which in turn means that the vanadium carbide and nitride will be more finely dispersed.

Strength is controlled by adjusting the cooling after hot forging. If cooling is not controlled accurately, this is likely to cause a large dispersion in strength. An automatic forging system and a special cooling hanger are normally used to control cooling. Final strength is also very sensitive to the chemical composition of the steel, and this must be adjusted carefully.

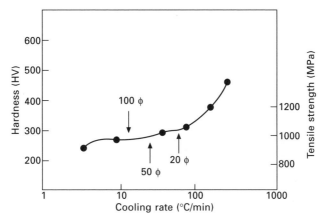

8.26 Relation between hardness and cooling rate of a micro-alloyed steel. The figure typically indicates the cooling rates for rod diameters of 100, 50 and 20 mm. The microstructure becomes finer as the cooling rate is faster.

The use of lead-free micro-alloyed steel for crankshafts has been suggested for environmental considerations.[28] Conventional micro-alloyed steel contains Pb (typically, Fe-0.45%C-0.26Si-0.8Mn-0.019P-0.023S-0.1V-0.16Pb), whereas lead-free steel has a chemical composition of Fe-0.45%C-0.01Si-1.12Mn-0.017P-0.151S-0.1V. The inclusion of MnS gives good chip breakability.

In early types of micro-alloyed steel, impact strength was low due to the coarse grain size that resulted from the slow cooling process. For crankshafts, impact strength is not so important, but it is crucial for suspension parts. Since 1985, improvements in toughness have been achieved without reducing machinability. Figure 8.27 shows how strength and toughness of micro-alloyed steel developed over time.[29] The original micro-alloyed steel had a medium carbon concentration and added V using precipitation hardening. The coarse ferrite-pearlite microstructure generated by slow cooling after hot forging, however, did not provide high toughness and as a result, the steel had a limited application.

High strength can be obtained without reducing toughness by reducing carbon and compensating for the resultant loss of strength by adding alloyed elements. This type of alloy generates bainite or martensite, but these microstructures are unstable in air cooling. Without appreciably changing the chemical composition and ferrite-pearlite microstructure, both strength and toughness are increased only by grain size refinement. As shown in Fig. 8.28,[30] grain size is reduced by controlling forging conditions and by adjusting steel quality. Forging at low temperature can reduce grain size, while the increased forging load shortens die life.

Another way to obtain fine grain size is to use inclusions in steel. Precipitated nitride and sulfide, such as TiN and MnS, can make the austenite grain fine

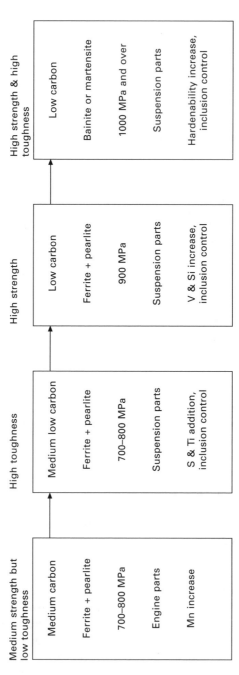

Medium strength but low toughness

Medium carbon

Ferrite + pearlite

700–800 MPa

Engine parts

Mn increase

High toughness

Medium low carbon

Ferrite + pearlite

700–800 MPa

Suspension parts

S & Ti addition, inclusion control

High strength

Low carbon

Ferrite + pearlite

900 MPa

Suspension parts

V & Si increase, inclusion control

High strength & high toughness

Low carbon

Bainite or martensite

1000 MPa and over

Suspension parts

Hardenability increase, inclusion control

8.27 Improvement of micro-alloyed steel.

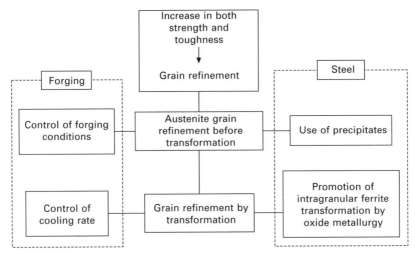

8.28 The methods to raise strength and toughness of ferrite-pearlite micro-alloyed steel. The transformation generates ferrite from austenite upon cooling. The oxide metallurgy necessitates a metallurgical technique utilizing fine oxide and sulfide particles to improve steel properties.

during forging. In addition, the precipitates act as nuclei for the ferrite to promote refining of grains on cooling after forging. The refined ferrite-pearlite microstructure raises toughness as well as strength, widening the application of this type of steel. A typical chemical composition is Fe-0.23%C-0.25Si-1.5Mn-0.03S-0.3Cr-0.1V-0.01Ti. The microstructure keeps the machinability high. Figure 8.29[31] shows toughness (impact value) and tensile strength of micro-alloyed steels.

The ferrite-pearlite microstrcture is successful below 1 GPa, but cannot generate strength above 1 GPa. For these conditions, a micro-alloyed steel with a bainite microstructure has been developed (Fig. 8.29).[31, 32] This alloy has higher Mn and Cr content with a small amount of Mo and B, so that it creates a stable bainite microstructure in air cooling. A typical chemical composition is Fe-0.21%C-1.5Si-2.5Mn-0.05S-0.3Cr-0.15V-0.02Ti.

There is still a need to develop strong but sufficiently machinable steel. Yield strength directly relates to fatigue strength and buckling strength. The higher the yield strength, the higher the fatigue and buckling strengths. On the other hand, machinability relates to the hardness. The higher the hardness, the lower the machinability. Hardness is proportional to tensile strength (σ_{UTS}), so machinability decreases with increasing σ_{UTS} of the steel.

In order to increase fatigue strength without reducing machinability, yield strength should be increased without raising the ultimate tensile strength. The ratio of yield strength to tensile strength, Ry, is given by σ_y / σ_{UTS}; a strong but machinable steel should have a high yield ratio value.

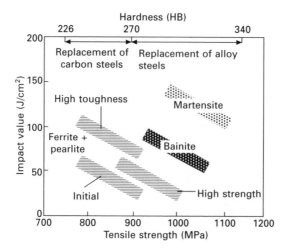

8.29 Impact value vs. tensile strength for microstructures of micro–alloyed steel.

In comparison with normal carbon steel after normalizing or toughening, micro-alloyed steel generally has a lower yield ratio. Typically, the Ry value of a normalized steel measures around 0.85, while that of a micro-alloyed steel is around 0.7. Changing the chemical composition can improve the value. The yield ratio is an important indicator in developing a well-balanced micro-alloyed steel.

Micro-alloyed steel does not need conventional quenching and tempering, therefore costs are lower and the steel is suitable for intricate shapes, because the thermal distortion that accompanies quench-hardening is avoided.

Micro-alloyed steels can give good strength in the as-rolled condition, after forging or cold working, and their use for automotive steel parts is increasing. High-strength bolts are made of high-strength micro-alloyed steel that has improved cold forgeability. Micro-alloyed steel for cold heading wire rod is increasingly used at tensile strengths above 800 MPa.[33]

8.7   Strengthening

The crankshaft is forced to work under a repetitive load. Figure 8.30 shows a fatigue fracture at the shaft portion of an assembled crankshaft which initiated from a non-metallic inclusion in the steel. Figure 8.31 shows a fracture observed in a test specimen after fatigue testing. It shows fatigue failure caused by an inclusion below the surface, where the crack initiated at the inclusion has spread to the surface and resulted in failure.

Without such inclusions, fatigue strength fundamentally depends on the strength of the material. In the crankshaft, stress concentrates at the corner

8.30 Fatigue fracture of a carbon steel S50C crankshaft. The side view (broken at the left-hand end) is on the right. An inclusion was the starting point of the crack. A typical beach mark initiated at a shallow position from the surface is observable (upper right in the left photo). This is a rare example because recent refining technology has drastically decreased the number of nonmetallic inclusions.

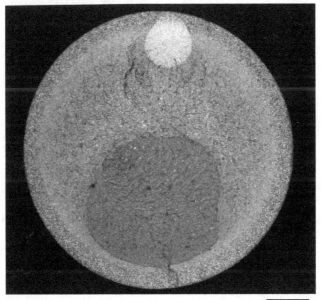

1 mm

8.31 Fatigue fracture observed in a bar test piece (carburized Cr-Mo steel SCM420). The upper round area indicates crack initiation. The crack initiated at an inclusion below the surface. The inclusion was observed at the center of this round area.

radius between the crankpin and web, and at the oil hole or keyway of the shaft. A sharp edge and rough surface are likely to concentrate stress. In actual parts, cracking often initiates as a result of the surface shape. Forging defects in shaped material such as forging laps (folds) must be avoided. Figure 8.32 shows a fatigue fracture initiated by fretting wear of the crankshaft end portion, where the flywheel magnet is force-fitted (the right end in Fig. 8.2).

20 mm

8.32 Fatigue fracture at the fit portion of the flywheel magnet. The fretting wear at the surface has initiated the crack.

Stress analysis using the finite element method predicts fatigue strength relative to shape. A low-weight part is designed and tested,[34] and if it breaks during testing, durability is improved by slightly increasing the thickness of the crankshaft. The part is tested again and the process repeated, until it meets the strength requirements.

Simulation testing, which reproduces the actual stress seen during operation, is carried out in some cases. Figure 8.33 represents schematically a fatigue-testing machine that uses resonance. The vibrator applies a vibrational load (the arrows) to the crankshaft webs and the feedback from a strain gauge attached to the surface is used to control the applied stress. The stress at the corner radius between the crankpin and web (where fatigue failure usually takes place) is controlled. Figure 8.34 compares the fatigue strength of identically shaped crankshafts, measured by this testing machine. The use of different materials and heat treatments demonstrates that fatigue strength is greatly influenced by them.

It should be noted that the strongest material is not always the most appropriate material to use. SCM 435L is the strongest, but has low

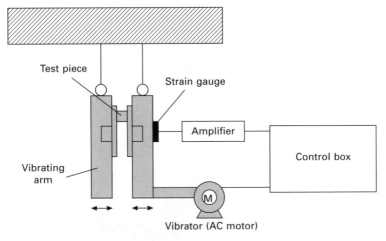

8.33 Resonance type fatigue-testing machine. Vibrating arms connected to the vibrator are attached at the web of the hanging crankshaft.

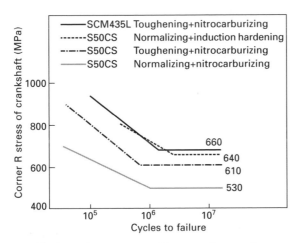

8.34 Fatigue data measured by actual parts. L means leaded free-cutting steel and S sulfured free-cutting steel.

machinability due to its high hardness, and when the pin diameter of the crankshaft is thin, distortion is likely to occur in the specification of the induction hardened S50C.

Microstructural control of fiber flow is another important aspect that must be considered. Figure 8.35 shows the cross-section of a forged gear, in which the linear microstructure looks like a fiber. The contrast in fiber flow is due to the layered distribution of ferrite and pearlite. After chemical etching, the difference in corrosion resistance of both microstructures exposes the fiber-

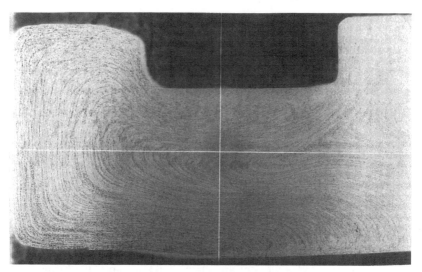

8.35 Fiber flow in a cross-section of a gear. The central cross line is for measuring. The fiber is observable by the inhomogeneous distribution of chemical composition elongating toward the extended direction.

like pattern. The flow originates from local inhomogeneity of chemical composition (segregation) generated during casting, and this segregation in the cast ingot elongates longitudinally during the shaping process. Figure 8.36(a) illustrates the fiber flow of the original billet. Forging shapes the fiber flow, as shown in Fig. 8.36(b), and the annual ring-like fiber flow in Fig. 8.35 is formed in this way.

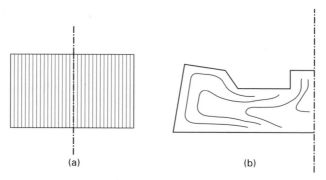

(a)   (b)

8.36 (a) Fiber flow of a billet cross-section. (b) Schematic illustration of Fig. 8.35.

Impact strength parallel to the fiber direction is double that for a perpendicular impact, so fiber direction is important, particularly for parts used under shock loading. Figure 8.37(a)[35] schematically illustrates the fiber flow in a forged crankshaft and Figure 8.37(b) shows that in a crankshaft machined from a bar. The forged type is recommended. For small production numbers or a prototype part, machining from a bar is the more common method, but the part is weaker than a forged part.

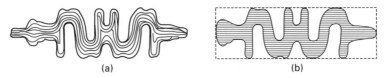

(a)   (b)

8.37 (a) Fiber flow of a forged crankshaft. (b) Machined crankshaft from a bar (broken line).

Deep rolling is used to strengthen the fillet radii between crankpin and web. The slight surface deformation at this point increases high compressive residual stress in a similar way to shot peening, and generates resistance to fatigue failure. Deep rolling is carried out using a small roller during the machining process.

Fatigue strength is very important, but productivity considerations must take machinability and forgeability into account. Generally, cheap, normalized

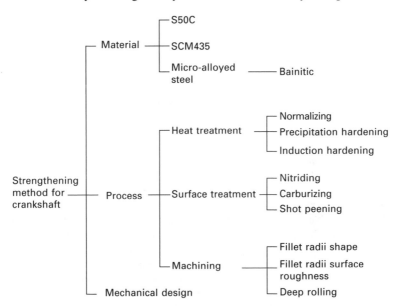

8.38 Methods to strengthen crankshafts. The material selection, production process and excellent mechanical design improve the strength. Deep rolling is a treatment raising fatigue strength through work hardening. The roller-like knurling tool plastically deforms the fillet radii.

carbon-steel is used, and if the strength is insufficient, nitrocarburizing is carried out. For more demanding requirements, an alternative steel is chosen, using data such as that given in Fig. 8.34 to support the choice.

The forged crankshaft requires considerable machining due to its complicated shape, and cast iron with its high machinability is an attractive material for many low and medium output power engines. Nodular cast iron is used most often because of its high strength, and deep rolling and nitrocarburizing are frequently used to improve fatigue strength. Figure 8.38 summarizes the different measures used to strengthen the crankshaft.[36]

8.8   Conclusions

A crankshaft is the heaviest moving part in the engine. Since it works as a rapidly moving weight, light materials are not suitable. Nitrocarburizing, induction hardening and carburizing are three of the most frequently used case-hardening methods, and give compressive residual stress to the surface, thereby significantly improving fatigue strength.

8.9   References and notes

1. Unlike cast iron, the swarf is likely to tangle in the cutting tool and this lowers productivity. The best machining condition is selected in order not to cause such a problem.
2. The assembled crankshaft for a two-stroke engine is designed by taking its volume into consideration, because the engine constitutionally compresses the combustion gas in the crankcase. To get a high compression ratio, the dead space volume in the crankcase should be minimized.
3. Hayashi Y., *Reesuyou NA Engine*, Tokyo, Grand Prix Publisher, (1993) 125 (in Japanese).
4. *Forging Handbook*, Forging Handbook-Editing Committee, Tokyo, Nikkan Kougyou Shinbunsha, (1971) 7 (in Japanese).
5. Sakai T., *Nippon Kinzoku Gakkai Kaihou*, 22 (1983) 1036 (in Japanese).
6. *Saishin Soseikakou Youran*, Nippon Soseikakou Kyoukai, (1986) 194 (in Japanese).
7. *Keikinzoku Tanzou Techou*, ed. by Tanzou Techou Bukai, (1995) (in Japanese).
8. Hot forging heats the billet first. Then, the forging proceeds continuously through drawing, blocking, finishing and deburring. Reheating is not generally implemented at each stage. In hammer forging, the operator transfers and revolves the heated material manually. Even a complicated form can be shaped with a small number of dies. If a skilled worker can be hired, this is appropriate for a small lot size. By contrast, in press forging, the operator does not carry out skilled work like that in hammer forging, they just transfer the workpiece, so that each stamp needs a different die. For example, the connecting rod requires one shaping die and one trimming die in hammer forging. By contrast, it requires more than twenty dies in press forging, and the press machine has several dies installed in one platen. The workpiece is transferred automatically. Accordingly, it is expensive for production runs of less than 100,000.

9. Ochi T., *et al.*, *Nippon steel technical report*, 80 (1999) 19.

10. Effective case depth: the distance from the surface to the position of the Vickers hardness value of 550 HV. Yet, in this definition, there is a possibility that the whole portion of hard SCM 435 is counted as the case depth. In such a case, a proper hardness is prescribed. Total case depth: the distance (JIS G0557) from the surface to the position where the physical (hardness) or chemical (macrostructure) property becomes the same as the matrix. In Fig. 8.17, the effective case depth measures 0.9 mm, while the total case depth 1.3 mm. The total case depth D (mm) at time t (h) is predicted by the equation $D = K \cdot \sqrt{t}$, where K is a constant; 0.475 at 871 °C, 0.535 at 899 °C and 0.635 at 927 °C.

11. *Netsushori Gijutsu Binran*, ed. by Nihon Netsushori Gijutsu Kyoukai, Tokyo, Nikkan Kougyou Shinbunsha Publishing, (2000) 66 (in Japanese).

12. There are other methods such as pack carburizing, liquid (bath) carburizing or vacuum carburizing. They have rather low productivity.

13. Straightening of distorted carburized parts should be avoided because it frequently causes fine cracks in the surface.

14. ALD Vacuum Technologies AG., Homepage, http://www.ald-vt.com, (2003).

15. Kowalewski J., SECO/WARWICK Corporation, Homepage, http://www.secowarwick.com, (2003).

16. Naitou T., *Tekkouzairyouwo Ikasu Netsushorigijutsu*, ed. by Ohwaku S., Tokyo, Agune Publishing, (1982) 27 (in Japanese).

17. Maki M., *Sanyo Technical Report*, 2(1995) 2 (in Japanese).

18. *Kinzoku Netsushori Gijutsu Binran*, ed. by Asada H., *et al.*, Tokyo, Nikkan Kougyou Shinbunsha, (1961) 212 (in Japanese).

19. Liquid carburizing is implemented at temperatures above A_1 in a molten cyanide bath. The carbon diffuses from the bath into the metal and produces a case comparable with one resulting from gas carbonitriding in an atmosphere containing some ammonia. The composition of the case produced distinguishs it from cyaniding. The cyaniding case is higher in nitrogen and lower in carbon. Liquid nitriding employs the same temperature range as for gas nitriding. As in liquid carburizing and cyaniding, the case hardening medium is molten cyanide. Liquid nitriding adds more nitrogen and less carbon to the steel than do cyaniding and carburizing in cyanide bath.

20. Cr-Mo steels do not contain sufficient alloying elements for nitriding, so that the steels cannot generate sufficient hardness. However, the fatigue strength and wear resistance are improved.

21. Nitriding is carried out in the temperature range where α-iron exists. At a higher temperature, despite the high diffusion rate of N, nitrides are difficult to form. By contrast, carburizing is carried out at the austenite region. It uses the property that much carbon dissolves into austenite.

22. The Japanese sword has a beautiful curve. It has a composite structure, comprising both inner low carbon portion and outer high carbon portion. The quench hardening mainly takes place at the edge. This introduces an additional curvature to the sword. The curvature depends on the strain during quenching. The resultant subtle curvature is unpredictable, but it should be adjusted at proper value. If it is carried out seriously, even heat treatment becomes an art. Inoue T., *Materials Science Research Int.*, 3, (1997) 193.

23. Wtanabe Y., Narita N. and Murakami Y., *Nissan Gihou*, 50(2002)68.

24. Yamagata H., *et al.*, SAE paper 2003-01-0916.

25. Loveless D., *et al.*, *Heat treatment of metals*, 2(2001)27. This book summarizes induction hardening and heating. Rudnev V., *et al.*: *Handbook of Induction Heating*, New York, Marcel Dekker, Inc., (2003).

26. Induction hardening also defines effective case depth and total case depth. The effective case depth is larger for a higher carbon content. For example, the values for S45C and S50C are defined as the distance having the hardness values above 450 HV (JIS 0559). In Fig. 8.23, the effective case depth measures 0.7 mm and the total case depth 1 mm.

27. Suto H., *Kikaizairyougaku*, Tokyo, Corona Publishing, (1985) 110 (in Japanese).

28. Hashimura M., *et al.* *Shinnitesu gihou*, 378(2003) 68 (in Japanese).

29. Nishida K. and Sato T., *Sumitomokinzoku*, 48(1996) 35 (in Japanese).

30. Takada H. and Koyasu Y., Nippon steel technical report, 64(1995) 7.

31. Ikeda M. and Anan G., R&D Kobe steel engineering report, 52(2002) 47 (in Japanese).

32. Sato S., Sanyo technical report, 8(2001)68 (in Japanese).

33. Kaiso M. and Chiba M., R&D Kobe steel engineering reports, 52(2002) 52 (in Japanese).

34. To develop a new engine having a large displacement volume, the bore of a small engine is increased. This increase can raise output power while retaining the basic layout of the previous small engine. The crankcase can be used with only an additional small adjustment. This method can develop a light compact engine, although it raises the stress in the materials. Additional surface treatment or material change, etc., should overcome the problems.

35. *Tekkouzairyou Binran*, ed. by Sato T., Nippon Kinzoku Gakkai and Nippon Tekkou Kyoukai, Tokyo, Maruzen Co., Ltd., (1967) 334 (in Japanese).

36. The steps to strengthen the crankshaft manufactured by BMW are introduced. Conradt G., MTZ, 64(2003)135.

9
The connecting rod

9.1   Functions

Typical connecting rods are shown in Figs 9.1 and 9.2. The connecting rod is generally abbreviated to con-rod. The crankshaft con-rod mechanism transforms reciprocative motion to rotational motion. The con-rod connects the piston to the crankshaft to transfer combustion pressure to the crankpin. There are bearing portions at both ends, the piston side is called the small end, and the crankshaft side, the big end.

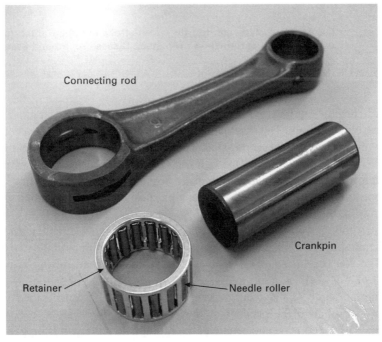

9.1 Monolithic con-rod. The lower left shows a needle roller bearing held in the retainer.

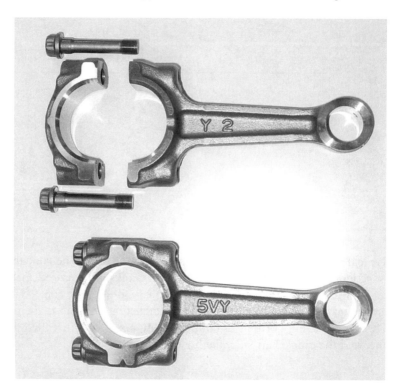

9.2 Assembly type con-rod for a four-stroke engine, a fracture-split con-rod using carburized Cr-Mo steel. Disassembled and assembled states.

The con-rod must withstand very high forces as the piston moves within the cylinder bore. The shaft portion of the con-rod is subjected to bending as well as tension and compression. The bearing portions receive load from the weight of the piston and the con-rod. To avoid failure of the bearings, the con-rod should be made as light as possible. To avoid buckling, the rod portion usually has an I-beam shape because of the high rigidity-to-weight ratio of this shape. Figure 9.3 shows the cross-section.

Although con-rods for both four-stroke and two-stroke engines have an I-beam shape, the thickness distribution is slightly different in the two engines. The four-stroke con-rod receives a large tensile load during the exhaust stroke as well as a compressive load during the combustion stroke. The inertial force of the reciprocating mass generates a tensile load which is proportional to the product of the piston assembly weight, reciprocating mass of the con-rod and square of the rotational velocity. It is bigger than the compressive load above a certain rotational speed.

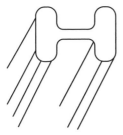

9.3 Section shape.

Figure 9.4 shows a fatigue fracture in an I-beam shaft. Beach marks typical of fatigue fracture are observable. The cracks initiated at the two corners as a result of bending caused by compression at the portion just below the small end. Although lighter designs are preferable, stress concentrations that can initiate fatigue failure must be avoided. There are two types of con-rod, monolithic and assembled. These types are used as shown in Table 9.1.

9.4 Fatigue fracture. The cracks initiated at the two corners and caused the beach marks.

9.2   The monolithic con-rod

Figure 9.1 shows a monolithic con-rod. The monolithic con-rod has a needle roller bearing at the big end, which is illustrated in Fig. 9.5. Single-cylinder and V-type twin-cylinder engines for motorcycles use monolithic con-rods. The two-stroke engine requires a needle roller bearing because the big end has less lubricating oil due to the structure. In four-stroke engines, lubricating oil is abundant in the crankcase, and the assembly type of con-rod is used because of the lower cost of this simpler structure.[1]

Figure 9.1 shows a needle roller bearing for the big end. The needle rollers held in the retainer are inserted into the big end and run on the outer

Table 9.1 Types of con-rods

Engine types	Two-stroke	Single-cylinder and V-type cylinder for four-stroke (mainly for motorcycles)	Multi-cylinder four-stroke
Monolithic type with needle roller bearing	x	x	–
Assembly type with plain bearing	–	–	x

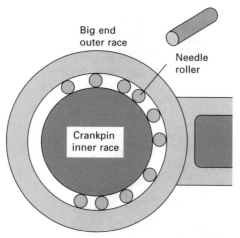

9.5 Big end boss of monolithic con-rod. Needle rollers are placed between the outer race and inner race (crankpin).

raceway of the big end and the inner raceway of the crankpin. The roller itself receives high stress and also exerts high Hertzian stress on the rolling surfaces. The retainer [2] (cage) separates the rollers, maintaining an even and consistent spacing during rotation, and also guides the rollers accurately in the raceways to prevent the rollers from falling out.

The big end is carburized to increase rolling contact fatigue strength, and honing finishes the surface accurately. Case-hardening steels such as JIS-SCM420 are used. Carburizing is required only at the rolling surface and copper plating is used as a coating to prevent other portions from carburizing. If carburizing hardens the entire con-rod, the subsequent straightening tends to cause cracking.

Figure 9.6 shows abnormal wear at a crankpin and Fig. 9.7 shows the counterpart of the big end. The causes of such abnormal wear include:

• inappropriate mechanical design, such as excessive loading, insufficient rigidity of the big end, low dimensional accuracy or insufficient lubrication

9.6 Wear at crankpin. The direct cause of this wear is that the needle rollers do not rotate in between the crankpin and con-rod.

9.7 Wear at big end.

- inappropriate composition of lubricating oil
- oil contamination by dust
- inappropriate material properties.[3]

Materials should not be used under excess loading. However, since newer engine models usually require higher power output, designs are likely to specify higher loading. Current technologies related to dimensional accuracy, material, heat treatment, tribology, etc., are used to solve the problems presented by the new designs.

9.3   The needle roller bearing

9.3.1   Fatigue failure

The needle roller bearing (Fig. 9.1) works under high bearing loads in a limited space in the big end. The rollers implement planetary motion between the crankpin and the big end, and the smaller diameter makes the big end light, thus lowering weight but at the same time increasing contact stress. Soft silver-plating protects the side surface of the retainer holding the rollers from side thrust.

The performance and life of bearings are very important in extending the life of the engine. Rolling contact fatigue is likely to take place at high-speed revolutions.[4] Fatigue failure of the pathway surface is called rolling contact fatigue failure. This occurs when the finished smooth surface breaks under repeated rolling. Generally, failure caused by high contact stress appearing in various morphologies: pitting or spalling shows small holes; wear cracking occurs at a right-angle to the sliding direction; flaking is accompanied by flaky wear debris; case-crushing takes place in case-hardened steel, and so on. Rolling contact fatigue life is significantly influenced by material factors such as carbide shape in steel and nonmetallic inclusions.

Table 9.2 shows the chemical composition of a bearing steel used for needle rollers, JIS-SUJ2. Figure 9.8 shows the microstructure of a needle roller containing spheroidized carbide. It is a hyper-eutectoid steel with hard carbide dispersed in a high carbon matrix (see Appendix F). A special heat treatment generates the fine round carbide. The finer the carbide, the longer the fatigue life. Hyper-eutectoid steel is slowly cooled after hot working and normally generates a mixed microstructure of lamellar pearlite in grains and net-shaped carbide at grain boundaries. Figure 9.9(a) illustrates this microstructure schematically. The net-shaped carbide is very brittle and such a microstructure is undesirable in high loading situations, however, spheroidizing generates fine spherical carbide through breaking the carbide net, thus improving brittleness.

Table 9.2 Chemical composition of bearing steel JIS-SUJ2 (%)

JIS	C	Si	Mn	P	Cr	Mo
SUJ2	1	0.2	<0.5	<0.025	1.5	<0.08

In order to produce spherical carbide, the carbide net must be fragmented,[5] in a process illustrated in Fig. 9.11(a). By keeping the steel in the austenitic temperature region above the Acm line (point (a) in Fig. 9.10), the carbide net at the grain boundaries dissolves into austenite. Lamellar pearlite dissolves simultaneously during this procedure.

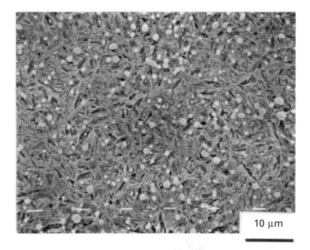

9.8 Microstructure of a needle roller under scanning electron microscopy. Spherical carbide around 2 μm disperses in tempered martensite matrix. The steel containing fine round carbide is ductile, although the carbide itself is hard and brittle.

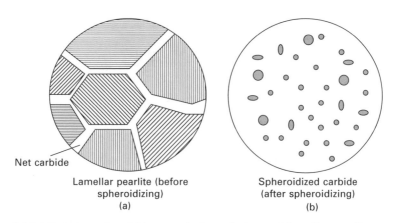

Net carbide

Lamellar pearlite (before
spheroidizing)
(a)

Spheroidized carbide
(after spheroidizing)
(b)

9.9 (a) Net-shaped carbide at grain boundaries and lamellar pearlite in grains of a hyper-eutectoid steel. The microstructure resembles the super-carburized microstructure shown in Fig. 8.15 of Chapter 8. (b) Spheroidized carbide (cementite).

The next stage is the spheroidizing process (Fig. 9.11(b)) at the mixed region of austenite and cementite (point (b) in Fig. 9.10, above the A_1 line). After this, slow cooling to below the A_1 line spheroidizes the carbide. In this procedure, round carbide is generated spontaneously because the spherical shape has less surface energy. A sufficient number of nuclei are required in order for fine carbide to be dispersed. If carbide nuclei are not present above A_1, the supersaturated carbon in the austenite generates lamellar pearlite during cooling to below A_1 and spheroidization fails.

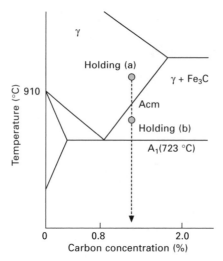

9.10 Austenite area in the iron-carbon phase diagram. The temperatures corresponding to (a) and (b) in Fig. 9.11 are indicated.

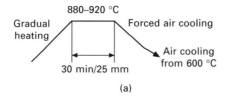

(a)

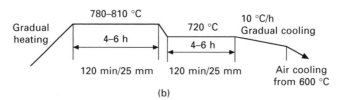

(b)

9.11 Spheroidizing diagram. (a) The process removes net carbide and refining lamellar pearlite. The representation 30 min/25 mm means that the treatment requires 30 min. for a 25 ϕmm rod. (b) Spheroidizing treatment. Additional quenching and tempering are necessary for a roller bearing.

The fine carbide in a homogeneous sorbite microstructure (see Appendix F) dissolves into the austenite above the A_1 point. However, carbide nuclei are eliminated if the temperature is too high or the time period too long. Conversely, if the temperature is too low or the time period too short, then an excessive number of nuclei form. In both cases, the desired amount of spheroidizing is not achieved. Temperature and time must be controlled accurately to produce the correct number of carbide nuclei. Spheroidizing results in a bearing steel with a ferrite matrix containing round carbide.

To perform as a bearing, spheroidized steel needs further heat treatment to increase hardness. The hardening process consists of oil quenching followed by tempering, which adjusts the hardness value to the range of 58–64 HRC. Both heating time and temperature before quenching are very important. During heating above Ac_1, carbide at about 5 vol. % dissolves into austenite, while the undissolved carbide remains.

If quenching is too slow or the temperature too high, the decomposition of carbide increases, raising the carbon concentration of the matrix and therefore increasing the amount of austenite retained in the quenched microstructure. This austenite is soft, and gradually transforms to martensite during operation, causing expansion of the bearing in a distortion that will eventually cause the bearing to fail. This must be balanced with the fact that an appropriate amount of retained austenite prolongs rolling contact fatigue life.

If heating is too short or the temperature too low, there is insufficient decomposition of carbide, which reduces hardness. The tempering temperature for a needle roller is set at 130–180 °C in order to generate slightly higher hardness. As an inner or outer ring, it is tempered at 150–200 °C. Carbonitriding is frequently used as an additional heat treatment before quenching because the nitrogen in the carbonitrided layer gives high wear resistance, particularly under contaminated lubricating oil. This process is carried out in the austenitic region (see Chapter 8).

A similar spheroidizing treatment is also carried out for low-carbon steel because steel with spheroidized carbide shows high malleability (see Appendix F). The lamellar pearlite changes to spheroidized pearlite and the finely distributed carbide in the soft ferrite matrix greatly raises cold forgeability.

9.3.2   Factors affecting the life of bearings

The carbide shape significantly influences fatigue life under rolling contact. In addition to this, the amount of inclusions in the steel also influence fatigue life.[6] Carbide works as a notch, causing microscopic stress concentration and initiating fatigue cracks. The inclusions originate from an involved slag (typically $MgO \cdot Al_2O_3 + CaO_n\, Al_2O_3$ generated in the steel-making process). On the one hand, the slag consists of glassy oxides that have low melting points and absorb harmful impurities from the molten steel during the refining process, but on the other hand, if it remains in the product, the inclusions have a detrimental effect. Figure 9.12[7] shows the effect of the refining process on the durability of bearing steel. This figure illustrates percent failure against life plotted as a Weibull distribution. The values on the vertical axis are typical in this field.

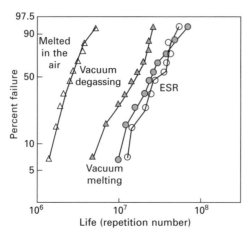

9.12 Effect of refining on the life of bearing steel.

Bearing life refers to the number of times any bearing will perform a specified operation before failure. It is commonly defined in terms of L10 life, which is sometimes referred to as B10. The bearing's L10 life is primarily a function of the load supported by (and/or applied to) the bearing and its operating speed. L10 life indicates the fatigue life by the repetition number at which 10% of the tested samples break. Alternatively, at L10, 90% of identical bearings subjected to identical usage applications and environments will attain (or surpass) this number of repetitions before the bearing material fails from fatigue.

Many factors influence the actual life of the bearing. Some of the mechanical factors are temperature, lubrication, improper care in mounting, contamination, misalignment and deformation. As a result of these factors, an estimated 95% of all failures are classified as premature bearing failures. Secondary refining removes inclusions from steel. In Fig. 9.12, the ESR material has the longest life. The left-hand line corresponds to the old technology, which does not include secondary refining. This diagram reflects the history of the refining technology of steel.

Shown in Fig. 9.13[8] is the relationship between rolling contact fatigue life L10 and the size of nonmetallic inclusions. As the size increases, the fatigue life becomes shorter. There are various types of inclusions, but it is known that the nonmetallic inclusions that reduce L10 life stem particularly from oxide. Figure 9.14 shows more clearly the relationship of L10 life to oxygen concentration.[8] The width shows the dispersion range and demonstrates that decreased oxygen content remarkably lengthens fatigue life. The size of nonmetallic inclusions relate to the oxygen content. The higher the oxygen content, the larger the size of the nonmetallic inclusions, and the shorter the fatigue life.

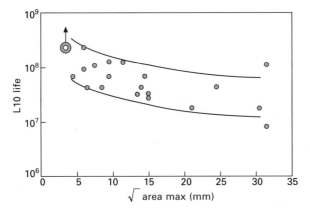

9.13 Relation between the size of nonmetallic inclusions and rolling contact fatigue life L10 of bearing steel JIS-SUJ2. The size is shown as the value √area max; the square root of the area of the biggest inclusion (area max). Reduced inclusion size logarithmically lengthens the life. The mark ⊚ indicates the desired value at present.

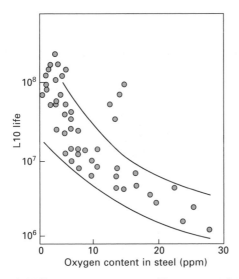

9.14 Oxygen content vs. rolling contact fatigue life L10 of JIS-SUJ2.

9.3.3   Secondary refining after steel-making

The increased life of bearing steel is due to improvements in the refining process. Refining is carried out in conventional steel-making, but secondary refining is necessary to reduce inclusions sufficiently to meet requirements. After primary refining, steel still contains nonmetallic inclusions such as Al_2O_3, MnS, $(Mn, Fe)O \cdot SiO_2$ and so on. These inclusions are internal defects and cause cracking. To obtain high-quality steel, molten steel must be refined further. The secondary refining process was developed following detailed research on the formation mechanism of nonmetallic inclusions (deoxidization, aggregation, and separation through surfacing), gas behavior in molten steel, the flow of nonmetallic inclusions and the deoxidization equilibrium. Figure 9.12 illustrates the history of the reduction of oxygen in steel. Figure 9.15[9] illustrates some typical secondary refining processes. The vacuum removes gases from molten steel, and bubbling argon gas through molten steel removes nonmetallic oxides.

After secondary refining, the steel is continuously cast into bars. The high-quality steel obtained by secondary refining has fewer inclusions and is called clean steel; it is increasingly used for bearing steel and case-hardening steel. Carburized clean steel shows superior properties as a con-rod material, having high rolling contact fatigue resistance. Clean steel also has superior cold formability, leading to a greater use of cold forging.

9.4   The assembled con-rod

9.4.1   Structure and material

Multi-cylinder engines for cars and motorcycles use assembled con-rods like that shown in Fig. 9.16. The big end consists of two parts. The bottom part is called the bearing cap, and this is bolted to the con-rod body. Honing finishes the assembled big end boss to an accurate circular shape. The mating planes of the cap and rod body should be finished accurately in advance because this influences the accuracy of the boss. The plain bearing is sandwiched between the crankpin and big end.

Hot forging shapes the assembled con-rod. Cr-Mo steel JIS-SCM435 or carbon steel JIS-S55C are generally used. Free-cutting steels are frequently used when high machinability is required. Toughening is a typical heat treatment for carbon steel. The recent tendency to pursue high strength at reduced weight has led to the use of carburized SCM420 as well, which is very effective if the con-rod is designed to receive high bending loads.

9.4.2   The con-rod bolt

Con-rod bolts and nuts clamp the bearing cap to the con-rod body, sandwiching the plain bearing (Fig. 9.17). The bolt is tightened with an appropriate load to prevent separation of the joint during operation, and so the bolt must be able to withstand the tightening load and the maximum inertial force.

To reduce the weight of the big end, the bolt hole should be positioned close to the big end boss. Some bolt heads have elliptical shapes to prevent them from coming loose. To prevent the joint between the cap and body from shifting, the intermediate shaft shape of the close-tolerance bolt should be

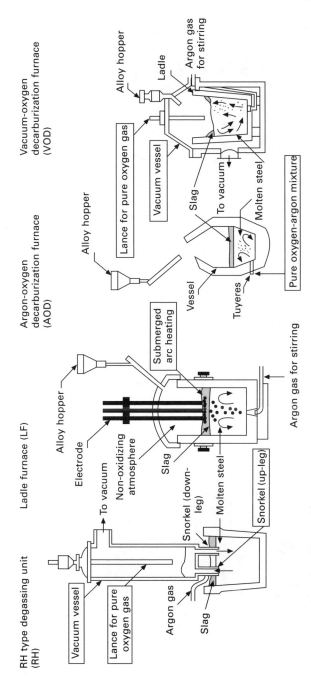

RH type degassing unit (RH)

Vacuum vessel

Lance for pure oxygen gas

To vacuum

Argon gas

Slag

Snorkel (down-leg)

Molten steel

Snorkel (up-leg)

Ladle furnace (LF)

Alloy hopper

Electrode

Non-oxidizing atmosphere

Slag

Submerged arc heating

Argon gas for stirring

Argon-oxygen decarburization furnace (AOD)

Alloy hopper

Lance for pure oxygen gas

Vessel

Tuyeres

Pure oxygen-argon mixture

Vacuum-oxygen decarburization furnace (VOD)

Alloy hopper

Ladle

Argon gas for stirring

Vacuum vessel

Slag

To vacuum

Molten steel

9.15 Secondary refining processes.

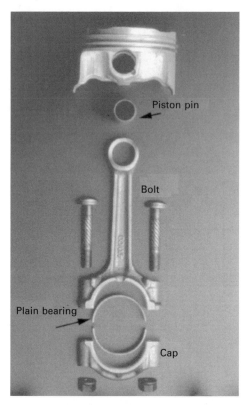

9.16 Con-rod big end and small end. The plain bearing is inserted at the big end.

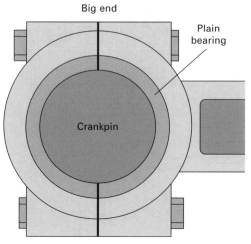

9.17 Big end boss of an assembly type con-rod. A pair of split plain bearings is placed on the crankpin.

finished accurately. The pitch of the screw portion must also be narrow. Thread rolling on toughened Cr-Mo steel SCM 435 is used to produce screws, and plastically shaped screws show very high fatigue strength.

The nut paired with a bolt is a separate part (Fig. 9.16). Some con-rods do not use a nut because the cap screw threads into the con-rod body itself. Figure 9.2 shows a con-rod screw that does not use nuts. This type can lighten the big end, but is likely to cause stress concentration on the screw thread. Using a nut can help to prevent fatigue failure in bolts.

The inertial forces from the piston, piston pin and con-rod body tend to separate the joint between the body and cap. Even a slight separation increases friction loss at the big-end boss, and shortens the life of the plain bearing. The stress on the con-rod bolt relates not only to the shape of the big-end boss but also to the rigidity of the bolt itself. The big-end boss should remain circular when the connecting rod bolts are tightened. The mating planes in the joint should lock the con-rod body and cap in perfect alignment, hence smooth mating surfaces are required. Stepped mating planes can prevent the joint from shifting. An additional method, fracture splitting, is discussed in Section 9.6, below.

Figure 9.18[10] shows distortions in the big-end bore under load. The con-rods under comparison have the same shape but are made of different materials; titanium (Ti-6Al-4V, indicated as TS) and Cr-Mo steel SCM435 (SS). Both circles show upward elongation, while the titanium con-rod, which has a lower Young's modulus, shows the larger distortion.

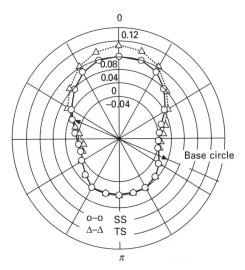

9.18 Roundness mismatch of the big end bore under loading.

9.5   The plain bearing

In the assembled con-rod, a plain bearing is generally used. The split plain bearing shown in Fig. 9.16 rides on the crankpin, fitting between the con-rod and the crankpin. It is a removable insert, as is the main bearing insert that supports the main journals of the crankshaft.

The crankpin rotates at a peripheral velocity of about 20 m/s. The piston and con-rod produce several tons of downward force. The plain bearing receives a contact pressure of typically around 30 MPa. The contact pressure is the pressure that the unit area of the sliding surface receives. The contact pressure (P) is calculated with the load (W), the shaft diameter (d), and the bearing width (L). $P = W/(d \times L)$. An appropriate gap is necessary between the plain bearing and crankpin so that oil penetrates the gap to lift up the crankpin, providing hydrodynamic lubrication during rotation. The plain bearing must conform to the irregularities of the journal surface of the crankpin. It should have adequate wear resistance at the running-in stage, high fatigue strength at high pressure and sufficient seizure resistance at boundary lubrication.

The plain bearing should also have the ability to absorb dirt, metal or other hard particles that are sometimes carried into the bearings. The bearing should allow the particles to sink beneath the surface and into the bearing material. This will prevent them from scratching, wearing and damaging the pin surface. Corrosion resistance is also required because the bearing must resist corrosion from acid, water and other impurities in the engine oil.

In the 1920s, plain bearings used white metal (Sn-Pb alloy). The allowable contact pressure was only 10 MPa. Because of this low contact pressure, the crankpin diameter had to be increased to decrease the contact pressure. To overcome this, a Cu-Pb alloy bearing having a higher allowable contact pressure was invented. Ag-Pb alloy was invented towards the end of the 1930s, and indium overlay plating of the Ag-Pb bearing was introduced during the Second World War. These important inventions enabled the plain bearing to work at an allowable contact pressure of up to 50 MPa. Recent advances have raised the allowable contact pressure to around 130 MPa. At present, two soft materials are typically used; Al-Sn-Si alloy[11] and Cu-Pb alloy. The Cu-Pb alloy is used for heavy-duty operations, such as diesel engines and motorcycles, and is capable of withstanding contact pressures over 100 MPa.

Figure 9.19 schematically illustrates the cross-sectional view of a plain bearing. It comprises three layers; the backing metal, which is a steel plate facing the con-rod, an intermediate aluminum alloy layer (Al-Sn-Si alloy) that has particulate Sn dispersed in the aluminum-silicon matrix, and a soft layer (Sn plating), called overlay, on the inside. The steel backing plate supports the soft aluminum alloy and the additional soft overlay gives wear resistance during running-in.

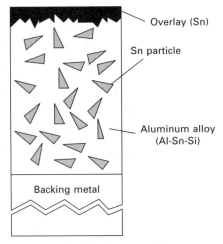

9.19 Cross-section of a plain bearing consisting of three layers.

The Sn overlay has a low melting temperature of 232 °C. Friction heat is likely to accelerate the diffusion of Sn into the bearing layer and cause a loss of Sn from the overlay. To prevent this, a thin layer of Ni is inserted between the bearing metal and overlay. This is referred to as the Ni dam.

The steel backing plate is laminated to the aluminum alloy sheet by cold rolling. Figure 9.20 illustrates schematically the rolling process. The high pressure between the rollers causes plastic deformation at the interface between the metals, resulting in strong metallic bonding. This two-metal structure is called clad metal. The plain bearing is shaped from the clad metal by a press machine. The Cu-Pb plain bearing also has a bimetal structure, where sintering laminates the Cu-Pb layer to the steel backing plate. In this process, a Cu-Pb alloy powder is spread onto the Cu-plated steel plate. The powder layer is sintered and diffusion-bonded to the steel plate at high temperatures.

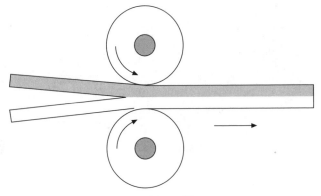

9.20 Bonding of clad metal by rolling.

Bearing metals contain soft metals such as tin or lead. These soft metals can deform to the shape of the adjacent part (the crankpin in this instance) and also create fine oil pools at the rubbing surface. However, if the bearing consists only of soft metals, it wears out quickly. Appropriate wear properties are provided by small particles of tin dispersed in the harder aluminum alloy. Lead particles perform this function in the copper-lead alloy. Although bearings containing lead have superior material properties, environmental considerations have led to the development of a Cu-Sn-Ag bearing[12] as an alternative to the Cu-Pb bearing. An Al-Sn-Si alloy is also being used to replace Al-Pb bearings.[13] The use of lead as a bearing metal is decreasing.[14]

9.6   Fracture splitting

As discussed above, the assembled con-rod uses bolts to fasten the bearing cap to the body (Fig. 9.16). The mating planes for the joint should be smoothly machined to lock the con-rod body and cap in perfect alignment. Positioning using a step mating plane or a knock pin, which prevents the joint from shifting, is sometimes used. These joint structures give good accuracy for plain bearings, but the machining required raises the cost of production.

To address this increase in cost, an alternative method using a broken jagged surface at the joint plane has been introduced. The con-rod in this instance is referred to as a fracture split con-rod, and Fig. 9.21 shows it in the dismantled state. The cap is cracked off to produce a rough mating surface as shown in Fig. 9.22. This surface helps lock the con-rod body and cap in perfect alignment and prevents the cap from shifting. The manufacturing process is as follows: first, forging and machining shape the big end monolithically. After completion, the monolithic big end is broken into two pieces (the bearing cap and the rod body), by introducing a crack at the joint surface. Special splitting tools have been developed in order to split the big

9.21 Fracture split con-rod (broken jagged con-rod) and the cap (right).

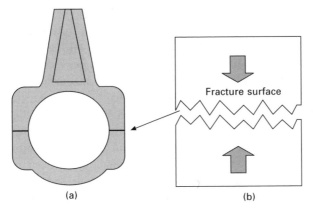

9.22 (a) Assembled state of the big end. (b) Schematic illustration of the broken mating planes.

end with minimal plastic deformation.[15, 16] To generate cracking at the correct position, notches are carved in the internal surface of the big end by laser or mechanical broaching. The fractured surfaces should fit exactly into position when both portions are overlapped and fastened by bolts. Any plastic deformation during splitting should be avoided, because if plastic deformation takes place, the broken surfaces cannot fit together. The crack should not cause branching, otherwise it is difficult to reassemble.

The fracture split con-rod is made from sintered steel,[17] hot forged high carbon steel or hot forged micro-alloyed steel.[18] Good mold yield of the sintering (powder metallurgy) method lowers costs. The manufacturing process for sintered steel has two steps, first, cold compaction of a powder in the mold and secondly, sintering the pre-form in a furnace to give enough bonding between powder particles. The typical chemical composition of sintered steel includes Fe-0.55% C-2.03Cu, and the microstructure shows ferrite and pearlite. The added Cu increases the density of the sintered part through liquid phase sintering. Additional hot forging (called powder forging) increases strength by removing small pores in the sintered steel. Splitting should take place in a brittle manner without plastic deformation, and the sintered con-rod is suitably brittle.

High carbon steel, around 0.7% C, is particularly good for this type of con-rod because it can be broken easily and the microstructure is pearlitic. Micro-alloyed steel with a typical chemical composition of Fe-0.7%C-0.2Si-0.5Mn-0.15Cr-0.04V has also been tested for fracture splitting. The vanadium is alloyed to give precipitation hardening properties, and the cooling process after hot forging is controlled so that precipitation guarantees strength.[18]

The use of fracture split con-rods is increasing because of low costs and high dimensional accuracy at the big end. To increase fatigue strength against bending, a case-hardened fracture split con-rod[19] has also been developed.

Figure 9.2 shows the con-rod for a motorcycle using Cr-Mo steel SCM420.

The fracture split con-rod was originally developed for outboard marine engines, which are placed at the stern of a boat. Two-stroke multi-cylinder engines are widely used because of good acceleration performance and high durability. The exhaust displacement measures about 3,000 cm³ at its maximum. The multi-cylinder engine employs a monolithic crankshaft because of the need for dimensional accuracy. However, a needle roller bearing is indispensable at the big end of the two-stroke engine, and so the con-rod should be of the assembled type. Since it is difficult to keep dimensional accuracy in the machined con-rod, fracture split technology has been introduced. The needle-bearing retainer also has to be split. Both ends require high hardness for needle rollers, so carburizing is used to give sufficient hardness to the rolling surface. The monolithic crankshaft is also carburized to improve wear resistance to needle rollers.

9.7   Conclusions

Lightweight con-rods can be made from Al-alloys or Ti-alloys. Low-power engines such as generators, can use a squeeze-cast Al con-rod, for which costs are low. Special engines, such as racing engines, use Ti alloy con-rods, which are light due to low specific gravity, but have medium fatigue strength.

The con-rod must be rigid. Both Al and Ti alloys are light, but the Young's modulus for both is low, so con-rods made from these metals must have a thick cross-section to give enough rigidity. The ratio of Young's modulus to density is nearly the same for steel, Al alloy and Ti alloy (= 2.6×10^7 Nmm/g). If this property is taken into account in the design process, then the Ti or Al alloy con-rod must be designed to weigh the same as a steel con-rod and so size becomes a consideration. Consequently, steel con-rods will remain the standard for the foreseeable future, and improvements in heat treatment and material characteristics will continue.

9.8   References and notes

1. The assembled crankshaft consists of several parts. A forging machine with a small capacity can forge these parts, resulting in lower costs than for the monolithic crankshaft, along with lower machining costs.
2. Without a retainer, the rollers of a needle bearing are likely to cause abnormal motion called skew, which leads to seizure or abnormal wear. The retainer maintains the intervals between the rollers to avoid this motion. Also, the roller does not have a perfect cylindrical shape but has a barrel shape, which prevents skew. The needle roller rolls with slip. The shaking of a crankpin and con-rod increases the slip. With sufficient lubricating oil, a plain bearing is suitable and costs less. Despite the low friction loss at the big end, no car engine presently uses a needle roller bearing for the big end.

3. *Junkatsu Handbook*, ed. by Japanese Tribology Association, Tokyo, Youkendou Publishing, (1987) 746 (in Japanese).
4. The small roller bearings have been improved by the development of high-power motorcycle engines. Okuse H., *et al.*: SETC Technical Paper 911279.
5. Koyanagi A., *Tekkouzairyouwoikasu Netsushori*, ed. by Oowaku S., Tokyo, Agune Publishing, (1982) 169 (in Japanese).
6. The defect size detrimental to rolling contact fatigue life is in the order of 10 μm. The critical size is smaller by one order than that required for a valve spring or a general machine part.
7. Fukuda K., *Tribologist*, 48(2003)165 (in Japanese).
8. Katou K., *Sanyo Technical Report*, 2 (1995) 15 (in Japanese).
9. *Tetsugadekirumade*, ed. by Nihon Tekkou Renmei, (1984) (in Japanese).
10. Tsuzuku H. and Tsuchida N., *Yamaha Gijutsukai Gihou*, 19 (1995) 20 (in Japanese).
11. Fukuoka T., *et al.*, SAE Paper 830308.
12. Kamiya S., *et al.*, *Tribologist*, 44(1999)728 (in Japanese).
13. Bierlein J.C. *et al.*, SAE Paper 690113.
14. Ito H., Kamiya S. and Kumada K. *Tribologist*, 48(2003)172 (in Japanese).
15. Weber M., SAE Paper 910157.
16. Catalogue of Alfing Kessler Sondermaschinen GmbH, (2000).
17. Mocarski S. and Hall D.W., SAE Paper 870133.
18. Park H., *et al.*, SAE Paper 2003-01-1309.
19. Kubota T., *et al.*, SAE Paper 2004-32-0064.

10

The catalyst

10.1   The development of catalysts for petrol engines

The air-polluting effects of internal-combustion engines were not recognized until the early 1960s. Up until that time, improvements in power output and exhaust noise were the main areas of development.[1] The driving force for change originated in the first measures to control air pollution, which were introduced in the smog-bound city of Los Angeles, USA. Controls for exhaust gases from motor vehicles were introduced in Japan and Europe soon afterwards. These early measures were focused on carbon monoxide (CO) and unburned hydrocarbon (HC).

The use of oxidizing catalysts to convert HC and CO has been mandated under exhaust gas regulations in the USA and Japan since 1975. The main components of the early catalysts were base metals such as Co, Cu, Fe, Ni, and Cr.[2] However, these were found to degrade over time, and precious metal catalysts were introduced to address problems of sulfur poisoning and metal evaporation. Unleaded petrol was developed because it was found that the lead in petrol coated the catalysts and made them ineffective.

An exhaust gas recirculation (EGR) system was introduced to decrease NOx emissions, but catalysts to remove NOx were not legally required until 1978. Regulations introduced in Japan (1978) and the USA (1981) required a further decrease in NOx emissions, and although the oxidizing catalyst system addressed HC and CO requirements, various controls in the engine were necessary to decrease NOx. As a result, power output fell and fuel consumption increased.

The first vehicle with a three-way catalyst was marketed in 1977, although it was not introduced for European cars until 1993. A system combining a three-way catalyst with electronic fuel injection (EFI) and oxygen sensors has now become the standard in petrol engines for cars. The three-way catalyst system reduces exhaust emissions after warming up, but recent legislation on emissions now requires a further decrease in pollutants, and reducing emissions at cold start is an important issue.

10.2   Structures and functions

Figure 10.1 lists the tasks for modern exhaust systems[1] and the various functions needed to bring about improved engine performance while keeping emissions low. The catalytic converter is an important component of the exhaust system and efforts to comply with emission regulations. The main pollutants are HC, CO, and NOx. Figure 10.2 illustrates the concentrations of these gases against air/fuel ratio. The concentration of each component varies with combustion, air/fuel ratio, EGR and ignition timing. HC derives

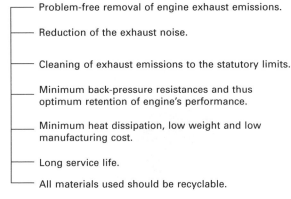

Problem-free removal of engine exhaust emissions.

Reduction of the exhaust noise.

Cleaning of exhaust emissions to the statutory limits.

Minimum back-pressure resistances and thus optimum retention of engine's performance.

Minimum heat dissipation, low weight and low manufacturing cost.

Long service life.

All materials used should be recyclable.

10.1 Tasks for modern exhaust system

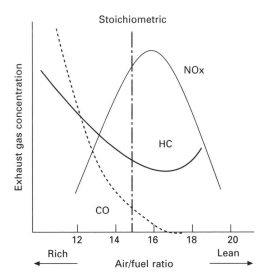

10.2 HC, CO and NOx concentrations in the exhaust gas of a petrol engine. The air/fuel ratio of 14.7 is called stoichiometric ratio. This is a theoretical ratio at which the fuel burns completely with air.

from unburned fuel. The concentration of HC decreases in lean combustion, but inversely increases in extremely lean combustion due to misfire. The concentration of CO does not depend on engine load, but does depend on the air/fuel ratio. The concentration of NOx is largely influenced by the air/fuel ratio and combustion temperature, and shows a maximum value at around an air/fuel ratio of 16.

Catalysts are materials that cause chemical changes without being a part of the chemical reaction. All exhaust gas must flow through the catalytic converter (Fig. 10.3a). The catalysts clean the exhaust gas by converting the pollutants to harmless substances, causing the reaction: $CO + HC + NOx \rightarrow CO_2 + H_2O + N_2$ inside the catalytic converter. The result is an exhaust gas containing less HC, CO and NOx. Normally, the complete unit is referred to as a catalytic converter, but strictly speaking, this term should only be used

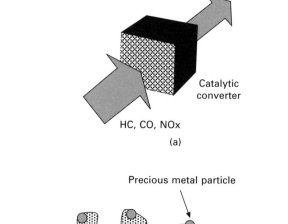

HC, CO, NOx

(a)

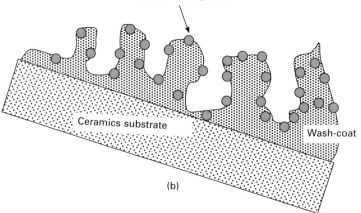

(b)

10.3 (a) Exhaust gas flows through the honeycomb. (b) Schematic illustration of the catalyst in the honeycomb cell. The rough wash-coat enlarges the surface area to hold the precious metal particles.

to describe the catalytic precious metals. These are platinum (Pt), rhodium (Rh) and palladium (Pd). Figure 10.3(b) schematically illustrates catalysts in a ceramic monolith. A honeycombed monolith of extruded ceramics (Fig. 10.4) or wrapped metal foils is normally used as the carrier, with the catalysts applied in a wash-coat covering the honeycomb substrate. With 62 honeycombs per square centimeter of flow area, the ceramic monolith offers a surface area of approximately 20,000 m^2 over which the exhaust gas can flow. This explains the cleaning effect of up to 98%, even though only 2 to 3 g of precious metals are used. The manufacturing process for the catalyst is shown in Fig. 10.7(c).

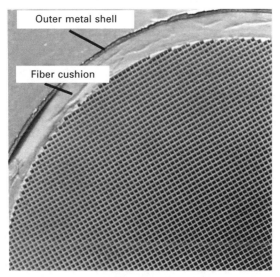

10.4 Quarter of a round ceramic monolith. A ceramic fiber cushion wraps the outer surface to prevent vibration of the metal casing.

Figure 10.5 illustrates the conversion characteristics of a three-way catalyst during exhaust gas purification. Pollutants behave very differently in the exhaust flow, as demonstrated by the NOx conversion in comparison to that of CO and HC. A common optimum for conversion of all pollutants has to be determined, and this is known as lambda (λ), or the lambda window. The highest conversion rate for all three components occurs in a small range around $\lambda = 1$. For the catalytic converter to be most effective, the air/fuel mixture must have a stoichiometric ratio of 14.7:1.

An oxygen sensor in the exhaust flow, the lambda sensor, controls the mixture electronically, keeping it at the optimum state over all engine loads. The oxygen sensor consists of a solid electrolyte, ZrO_2, which generates electromotive force (Vs) proportional to the oxygen concentration. The

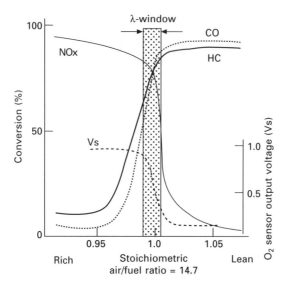

10.5 Characteristics of a three-way catalyst in exhaust gas purification. The three pollutants change with increasing air/fuel ratio. The output voltage of the O_2 sensor drastically changes around lambda = 1. This change can detect the air/fuel ratio and enables the sensor to control the stoichiometric combustion.

electromotive force of the sensor drastically decreases around the stoichiometric air/fuel ratio, $\lambda = 1$ (Fig. 10.5). This characteristic, combined with the EFI system, enables accurate fuel control. Without the sensor, EFI and control electronics, the three-way catalyst does not work well. Figure 10.6 illustrates the feedback control mechanism of the sensor and fuel injector.

Emission regulations for petrol engines in Europe, Japan and the USA are becoming increasingly restrictive. To meet future HC and CO limits and to improve fuel economy, manufacturers are looking forwards, running air/fuel ratios near lambda = 1 for full load engine conditions.

10.3   The three-way catalyst

10.3.1 Oxidation, reduction and three-way catalysts

Around 90% of all chemicals are manufactured using catalysts. Artificial catalysts are used in the manufacture of petrol, plastics, fertilizers, medicines and synthetic fibers for clothing. The word catalyst was first used by the Swedish chemist J. Berzelius and means 'to break down.'

A catalyst alters the speed of a chemical reaction but is left unchanged once the reaction has finished. For example, CO and O_2 do not react together at room temperature, and a mixture of these gases may remain stable for more than a thousand years if it is not heated. However, in the presence of a

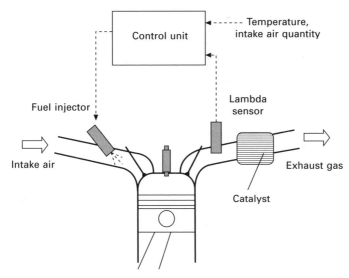

10.6 Feedback control of air/fuel ratio using a lambda sensor.

catalyst, the mixture rapidly changes to CO_2. During this reaction, the gas molecules are adsorbed onto the surface of the catalyst. This causes the bonding in the CO and O_2 molecules to relax, resulting in the atomic exchanges that form CO_2 and generating heat. The exhaust gas catalysts are functionally classified into three types, oxidizing, reducing and three-way. The oxidizing catalyst oxidizes HC and CO in an oxygen-rich atmosphere. The reducing catalyst reduces NOx even under oxygen-rich atmospheres – Cu/Zeorite is a typical example.

The first oxidizing catalyst for the petrol engine appeared in 1974, catalyzing the reaction between HC and CO in the presence of oxygen to form CO_2 and H_2O. The converter contained alumina pellets carrying the precious metal catalysts Pt and Pd. NOx was reduced by EGR or by adjusting combustion conditions. The EFI system enabled the oxidation catalyst to clean the exhaust gas at a lean air/fuel ratio range between 15.5 and 16.5. This drastically reduced the HC and CO concentrations, but because NOx was reduced by EGR or delayed ignition timing, the fuel consumption and drivability were not satisfactory.

Most catalysts used in petrol engines now are three-way catalysts that convert HC and CO into CO_2 and H_2O and reduce NOx to N_2. The catalyst comprises an alumina powder carrying Pt, Pd and Rh, with auxiliary catalyst CeO_2. Pt and Pd oxidize HC and CO, and Rh reduces NO. Rh works effectively even under low-oxygen conditions. Typically, the chemical reactions assisted by the components of three-way catalysts are:

$$C_3H_8 + O_2 \rightarrow CO_2 + H_2O \text{ by Pt and Pd}$$

$$CO + O_2 \rightarrow CO_2 \text{ by Pt and Pd,}$$

$$NO + C_3H_8 \rightarrow N_2 + CO_2 + H_2O \text{ by Rh}$$

Efficiency is influenced by several factors, including surface area of the catalyst and range of the lambda window. The available surface area of precious metal particles is maximized by using ultra small particles (1 nm) and dispersing them on the porous alumina substrate (Fig.10.3b). This basic technology was developed in the 1940s, when catalysts were used to increase the octane value of petrol.

The three pollutants are drastically reduced under conditions within the lambda window as shown in Fig. 10.5. However, small variations outside the lambda window increases exhaust emissions. The wider the lambda window, the wider the range of air/fuel ratios that the catalyst can clean. To widen the lambda window, CeO_2 is added as an auxiliary component. CeO_2 can store or supply oxygen via changes in its crystal lattice. Ce has two atomic values, Ce^{4+} or Ce^{3+}, and the valence number changes according to variations in the atmosphere, binding or releasing oxygen.

$$2Ce^{4+}O_2 \rightarrow Ce^{3+}{}_2O_3 + 1/2O_2.$$

This property compensates for deviations in the air/fuel ratio away from the stoichiometric ratio and therefore helps to maintain optimum conditions for catalytic conversion of the exhaust gases.

10.3.2 Deterioration of catalysts

There are three main causes for the deterioration of catalysts:

1. Physical failure due to thermal shock or mechanical vibration.
2. Poisoning by impurities such as Pb, P and S in the petrol and engine oil. and
3. Thermal failures such as sintering, where the precious metal and CeO_2 particles aggregate by diffusion and therefore reduce available surface area, and heating, which decreases the micro-pores in the alumina surface.

In order to prevent aggregation and the loss of Pd activity due to heat (because Pd operates effectively only at relatively low temperatures), a new catalyst[3,4] based on perovskite has been proposed. The perovskite-based catalyst, $LaFe_{0.57}Co_{0.38}Pd_{0.05}O_3$, has a self-regenerating property that preserves the catalytic function of Pd. As the catalyst cycles between the oxidizing and reducing atmospheres typically encountered in exhaust gas, Pd atoms move into and out of the perovskite lattice in a reversible process. Pd perovskite precipitates at the outside of the crystal during oxidation and dissolves to return into the perovskite crystal during reduction. This reversible movement suppresses the growth of metallic Pd particles, and hence maintains high catalytic activity during long-term use.

10.4   The honeycomb substrate

10.4.1 Ceramic

The monolithic honeycomb has replaced the pellet type as the most commonly used structure for catalytic converters. The ceramic monolith (Fig. 10.4) has proved to be an ideal carrier (substrate) for catalytic coatings containing precious metals. The honeycomb substrate must have the following properties:

- appropriate strength
- high heat and thermal shock resistance
- low back-pressure
- adequate adhesive strength to bond catalytic materials
- lack of chemical reactivity with the catalysts.

The starting materials for a ceramic honeycomb are magnesium oxide, alumina and silicon oxide, which are extruded and baked into cordierite ($2MgO \cdot 2Al_2O_3 \cdot 5SiO_2$). Figure 10.7(a) illustrates the manufacturing process of ceramic honeycomb. It has an optimal chemical resistance, low thermal expansion, high resistance to heat (melting point > 1400 °C) and can be recycled relatively easily. The standard monolith has a structure of 400 or 600 cpsi (cells/ inch^2). The fine honeycomb structure of the ceramic monolith calls for very careful embedding, or canning. Special covers consisting of high-temperature resistant ceramic fibers are used (Fig. 10.4). These insulate, protect and compensate for the different expansion coefficients of the monolith and steel casing.

10.4.2 Metal

Another type of honeycomb is made of metal foil. Figure 10.8 shows a typical metal honeycomb. Figure 10.7(b) illustrates the manufacturing process for a metallic honeycomb. The cell is constructed from a special, very thin and corrugated steel (typically, Fe-20%Cr-5Al-0.05Ti-0.08Ln-0.02C & N) foil.[5,6] Vacuum brazing using a filler metal such as Ni-19%Cr-10Si brazes the foil honeycomb directly into the steel casing.

The filler metal has a low Al content, so resistance to oxidation deteriorates near the bond. While the alumina protects the honeycomb from corrosion, it also obstructs bonding by diffusion at high temperature. A solid phase diffusion bonding method was developed to avoid the need for filler metal.[7] The alumina film on the foil surface is removed by evaporating aluminum atoms from the surface through vacuum treatment at high temperature, thus enabling diffusion bonding without the use of filler metal.

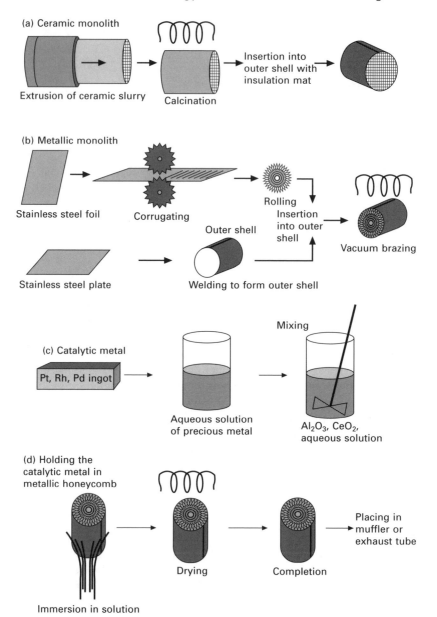

10.7 Manufacturing process of catalyst. (a) Ceramic monolith, (b) metallic monolith, (c) aqueous solution of catalytic metal and (d) putting the catalytic metal in the metallic honeycomb by immersion. The honeycomb is immersed in the aqueous solution made in the process (c). The water in the aqueous solution evaporates during drying process to precipitate the catalytic metal particles on the wash coat.

10.8 Metal honeycomb.

The metal honeycomb has a thin wall (40 μm), and therefore a lower back-pressure and a smaller construction volume for an identical surface area compared with the ceramic honeycomb. However, the metal honeycomb has several disadvantages, including higher costs, and higher radiation of heat and structure-borne noise, so additional insulation is required.

A wash-coat containing the catalyst covers the honeycomb substrate surface. γ–Al_2O_3 containing some auxiliary components is commonly used for the coating. It holds the precious metal particles and operates as an auxiliary catalyst. Figure 10.7(d) shows the process of coating catalytic metals on the honeycomb. The wash-coat should have the following characteristics:

- a large surface area to increase contact with the exhaust gas
- high heat resistance
- chemical stability against poisonous components
- lack of chemical reactivity with catalytic components
- adequate adhesive strength to bond to the substrate under high temperatures and drastic temperature changes.

10.5   The development of catalysts to reduce NOx

The need to decrease CO_2 while at the same time keeping fuel consumption low forces engines to operate under lean combustion conditions. Stable operation is now possible at an air/fuel ratio of 50. These conditions meant that the air/fuel ratio is beyond the lambda window, and the normal three-way catalyst cannot reduce NOx under such high oxygen concentrations.

Catalysts that reduce NOx under high oxygen concentrations are called lean NOx catalysts. Two types have been introduced, selective NOx reduction catalysts and NOx storage reduction catalysts. Selective NOx reduction catalysts include PT-Ir/ZSM-5[8] and Ir/BaSO₄,[9] and assist the reduction of NOx by HC in high-oxygen environments. Some have already been marketed, but further development is required.

The NOx storage reduction catalyst[10,11] stores NOx temporarily as a form of nitric acid salt NO_3^- (Fig. 10.9), reducing NOx in the exhaust gas. The NO_3^- adsorbents are alkali metals or alkaline-earth metals such as $BaCO_3$. If combustion takes place in the rich state with higher CO and HC, the accumulated NO_3^- is separated and reduced.

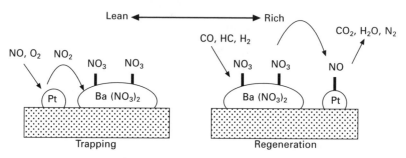

_10.9_ Mechanism showing trap and reduction of NOx.

The trapping process is:

$$NO + O_2 \rightarrow NO_2 \text{ and } BaO + NO_2 \rightarrow BaNO_3.$$

The regeneration process is:

$$BaNO_3 + CO \rightarrow BaO + N_2 + CO_2$$

The rich state occurs during acceleration or is generated by an intentional fuel control, the latter being known as rich spike. This system can serve to decrease fuel consumption and clean the exhaust gas, and was first marketed in a direct injection lean-burn engine by Toyota. One problem with this kind of catalyst is that the adsorbent also traps sulfur, and the sulfuric compounds decompose at higher temperatures than NOx. Accumulated S hinders the activity of adsorbents and shortens the life of the catalyst. Therefore, the sulfur content of the petrol must be kept as low as possible.

10.6   Controlling pollutants at cold start

Advances in emission control technology have succeeded in removing 100% of the regulated components after warming up. However, to decrease emissions further, the focus must now shift to emissions at cold start. The main cold start problem relates to the activation of the catalyst at low temperatures. The catalytic converter is a chemical reactor and the reaction rate mainly depends on the operating temperature. The catalyst does not work well in temperatures below 350 °C. Figure 10.10 lists some countermeasures.[12] Two technologies aimed at enhancing the activity of catalysts at cold start are discussed below.

10.6.1 Reducing heat mass and back-pressure

The stricter exhaust gas laws have raised demands on the monolith, requiring substrates with a larger surface area than the conventional 400 or 600 cpsi. The geometrical surface area of a substrate is mainly determined by cell density, while the wall thickness has very little influence. For an effective conversion rate, a high cell density is preferred. At a constant wall thickness, however, the mass of the substrate increases and the pressure drop increases due to a reduction in the open frontal surface area. The pressure drop obstructs the smooth flow of exhaust gas.

A high cell density thus increases the exhaust gas pressure drop and the thermal mass of the substrate. This can be partially compensated for by reducing the cell wall thickness, which in turn may influence the strength and durability of the substrate. Ultra-thin walled ceramic substrates with 900 and 1200 cpsi[13] and a wall thickness of between 2 and 2.5 mil (the unit mil represents 0.001 inch) have a high geometric surface area and a low mass. Figure 10.11[14] shows the light-off time (the time to the catalytic converter's effective phase) for HC and CO conversion as a function of cell density. Both heat up quickly and show good conversion behavior. The 900 cpsi/2 mil substrate is superior to the 1200 cpsi/2 mil substrate with regard to back-pressure and mechanical strength.

Thin-walled substrates with a high cell density have proven to be very effective for catalytic converters. They are lighter than the standard monolith, have a larger internal surface area and reach the catalytic converter's working temperature with a relatively low thermal input.

10.6.2 The close-coupled catalytic converter

The exhaust gas reaches temperatures of up to 900 °C very quickly after cold start. To use this energy to heat the catalyst, the converter has to be placed as close as possible to the engine. The exhaust gas in the exhaust pipe loses most of its heat energy in the first 1 m away from the engine. If the time

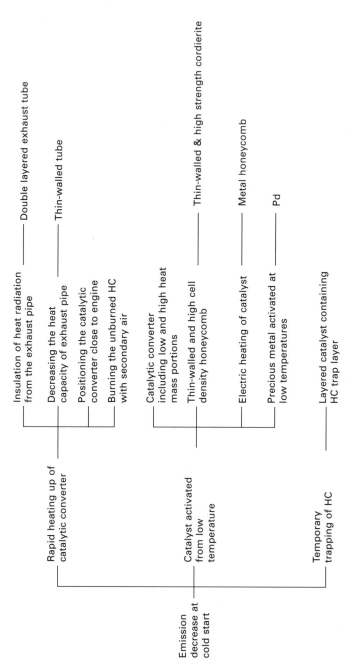

10.10 Methods to decrease emissions at cold start.

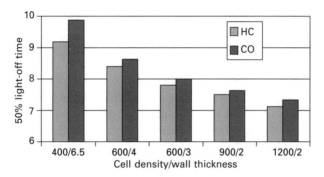

10.11 Light-off time for HC and CO conversion as a function of cell density.

between the catalytic converter's response and its effective phase is cut to around one quarter, the cleaning efficiency rises to almost 98%.

10.7   On-board diagnosis

As discussed above, the catalyst works best if combined with adjustments in engine operation. The functional reliability of the catalytic converter over the entire service life of a vehicle is of decisive importance for the lasting reduction of emissions. One possibility of ensuring this is on-board diagnosis (OBD), in which the vehicle computer continuously monitors the functional reliability of all components of the exhaust system. If a part fails or malfunctions, a signal lamp on the dashboard comes on and the error code is saved. In the case of the three-way catalytic converter, for example, the oxygen storage capacity of the catalytic converter, and thus indirectly the conversion itself, can be monitored. Signals from two lambda sensors, one in front and one behind the catalytic converter, are measured and compared, and the signal ratio is correlated with the degree of conversion for HC.

10.8   Exhaust gas after-treatment for diesel engines

10.8.1 Diesel particulate filters

Diesel engines are becoming more popular for cars in the European market, and this is encouraged not only by high performance combustion control but also by exhaust gas after-treatment. Basically, diesels are lean combustion engines, so NOx and particulates must be after-treated. The use of diesel engines in cars is expected to grow if particulates and NOx are well controlled. The relationship between the conversion efficiency of a three-way catalyst and air/fuel ratio is shown in Fig. 10.5. Petrol engines reduce NOx, HC and CO by controlling the stoichiometric air/fuel ratio. It is difficult to maintain stoichiometric combustion in a diesel engine, and therefore NOx cannot be reduced.

Particulate matter from diesel engines mainly consists of carbon microspheres (dry-soot) on which hydrocarbons, soluble organic fraction (SOF) and sulfates from the fuel and lubricant condense. The quantity and composition of the particles depends on the combustion process, quality of diesel fuel and efficiency of after-treatment. The soot is a solid and it is difficult to remove by catalysis. To decrease soot, fuel and air should be well mixed, but the resulting increased combustion temperature raises NOx. To decrease NOx, flame temperature is lowered using EGR or delayed injection timing. (Exhaust gas recirculation has been fitted to all light-duty diesels.) But this then results in an increase in soot and SOF, so a balance must be achieved between the amount of soot and the amount of NOx. Various technologies have been proposed to remove particulates from the exhaust gas. Oxidation catalysts are fitted to all new diesel-engined cars and will be fitted to light duty trucks. These oxidize the SOF and remove HC and CO, but cannot oxidize the soot.

Capturing particulates in a filter (diesel particulate filter DPF) is a solution. The filter captures all particle sizes emitted, but the problem is then how to eliminate the accumulated soot, which raises the back-pressure and could potentially cause a malfunction of the engine. The soot must therefore be captured and burned continuously in the filter. Soot burns in the region of 550 to 600 °C, but diesel car exhaust reaches only 150 °C in city traffic conditions. The problem of soot burn-off is referred to as regeneration.

Figure 10.12 shows a cutaway view of a typical DPF combined with an

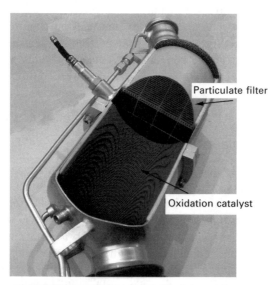

Particulate filter

Oxidation catalyst

10.12 DPF combined with oxidation catalyst.

oxidizing catalyst. The DPF has a different microstructure to the monolith for petrol engines. Figure 10.13 shows the mechanism. The channels in the DPF[15] ceramic monolith are blocked at alternate ends (Fig. 10.14). To pass through the monolith, the exhaust gas is forced to flow through the channel walls, which retain particulate matter in the form of soot but allow gaseous components to exit. This type of filter is called a wall-flow filter.

10.13 Mechanism of DPF.

10.14 DPF honeycomb.

The filter should be porous and should resist back-pressure. SiC is presently being used for car diesels, because it is more heat resistant and stronger than cordierite. The cheaper cordierite can be used if operational conditions are adjusted carefully on the combustion side and over-heating is avoided.

10.8.2 Regenerative methods

Regenerative methods fall essentially into two groups[16] as shown in Fig. 10.15.[17] Thermal regeneration raises the soot temperature to the light-off temperature by either electrical or burner heating, and catalytic regeneration chemically lowers the light-off temperature of soot. In thermal regeneration, the heater raises the temperature to burn away the soot. The thermal management of the filter during regeneration (temperature, oxygen content and flow rate) must be carefully matched to the requirements of the filter. Owing to fuel economy penalties incurred in thermal regeneration, these problems make thermal regeneration less attractive.

DPF type	System	Characteristics
Thermal regeneration	(1) DPF with electrical heater — Exhaust gas switching valve	Intermittent regeneration using bypass
Catalytic regeneration	(2) Fuel additive (Ce) — DPF (SiC) — Engine — Oxidizing catalyst	Fuel additive service system — Intermittent regeneration
	(3) NO → NO₂ (Oxidizing catalyst) — DPF	Increase of NO₃ conversion ratio — Continuous regeneration

10.15 Typical DPF technologies.

Catalytic regeneration is the alternative method. Soot burns in air at around 550 °C, while it will react with NO_2 below 300 °C. In the continuously regenerating trap (CRT), (3 in Fig. 10.15), the oxidizing catalyst placed before the DPF changes NO to NO_2. The NO_2 generated in this way continuously oxidizes and removes PM[16,18] through the reaction, $NO_2 + C \rightarrow NO + CO$.

The main obstacle to widespread introduction of the CRT is the effect of sulfur in fuel. The adsorption of SO_2 inhibits the adsorption of NO, hence blocking the formation of NO_2. This is common to all oxidation catalysis in diesel after-treatments. In this type of coated catalyst, the amount of S in the fuel must be low to avoid poisoning the catalyst.

10.8.3 Expendable catalyst additive

In 1999, PSA Peugeot Citroen successfully marketed[19] a DPF technology using an expendable catalyst additive and common rail fuel injection (2 in Fig. 10.15). The expendable cerium-based catalyst is added to the diesel fuel using an on-board container and a dosing system. The catalyst lowers the light-off temperature of soot to 450 °C. Combustion compensates for the residual temperature gap of 300°C (from 450°C to 150 °C). When soot accumulation in the filter becomes excessive, additional fuel controlled by injection raises the temperature of the soot. The rich exhaust gas from the engine also heats up the exhaust gas through an oxidation catalyst positioned before the particulate filter.

This system uses CeO_2 as the additive. The DPF filter is cleaned automatically every 400 to 500 km. A system that uses expendable additives does not depend on the sulfur level in diesel fuel. Various organic compounds are also known to have a catalytic effect for oxidizing particulates.[16]

10.8.4 The deNOx catalyst

The exhaust gas emitted by diesel and lean-burn petrol engines is comparatively rich in oxygen. This inherently facilitates the removal of HC, CO and PM through oxidizing reactions, but not the removal of NOx. Direct decomposition of NOx is too slow without a catalyst, so mechanisms using chemical reduction have been proposed. Figure 10.16[17] provides some typical deNOx mechanisms.

The NOx storage reduction type (1 in Fig. 10.16) is the same as that for

Type	System		Problems
NOx storage reduction	(1) Instantaneous rich state	Reduction by HC and CO	To obtain rich A/F ratio
Selective reduction	(2) Aqueous urea Catalyst	Reduction by NH$_3$	Urea service infrastructure Restriction of NH$_3$ slip
	(3) HC (fuel)	Reduction by HC	Increase of NO$_3$ conversion ratio

10.16 Typical deNOx technologies.

the gasoline engine (Fig. 10.9). The main problem is how to generate an instantaneous rich state. The catalyst also operates poorly with high-sulfur fuels. Selective reduction uses controlled injection of a reducing agent into the exhaust gas. DeNOx assisted by HCs (3 in Fig. 10.16) and urea (2 in Fig. 10.16) are currently being researched for diesel engines.

Ammonia is very effective at reducing NOx, but is toxic. An alternative is to inject urea, $((NH_2)_2CO)$, which undergoes thermal decomposition and hydrolysis in the exhaust stream to form ammonia.

$$(NH_2)_2\, CO \rightarrow NH_3 + HNCO$$

The NO and NO_2 reduction then proceeds with the assistance of a catalyst (e.g., $V_2O_5/WO_3/TiO_2$).

$$HNCO + H_2O \rightarrow NH_3 + CO_2$$

$$4NO + 4NH_3 + O_2 \rightarrow 4N_2 + 6H_2O \text{ and } 2NO_2 + 4NH_3 + O_2 \rightarrow 3N_2 + 6H_2O$$

This process is called selective catalytic reduction (SCR), and requires a metering system for injecting urea (as an aqueous solution). Fuel consumption does not increase because this method does not require excessive combustion control.

DPF is effective for particulate matter, and the deNox catalyst removes NOx. A system that enables simultaneous reduction of particulate matter and NOx has been proposed.[20] The DPNR (diesel particulate and NOx reduction system) combines a lean NOx trap catalyst with intermittent rich operation. The sulfur contained in diesel fuel causes damage to the catalyst itself, through the formation of sulfates, and the generation of SO_4^{2-}. Work is under way to reduce the S content of diesel fuel to below 10 ppm.

10.9   Conclusions

The new and more restrictive exhaust gas regulations have set a challenge for the treatment of exhaust gas. Emission limits can be reached or exceeded within a few seconds after an engine starts. Countermeasures include further reductions in crude engine emissions, a faster response time of the catalytic converter and an enlarged catalytic surface area. Further advances in catalytic converters, EFI and sensors now compete against efforts to develop electric vehicles and fuel cells.

10.10   References and notes

1. Ebespracher Co., Ltd, Catalogue, (2003).
2. Muraki H., *Engine technology*, 3(2001) 20 (in Japanese.)
3. Daihatsu, Homepage, http://www.daihatsu.com, (2002).
4. Nishihata Y., *et al.*, *Nature*, 418(2002)164.

5. Itoh I., *et al.*, Nippon steel technical report, 64(1995)69.
6. Hasuno S. and Satoh S., *Kawasakiseitetsu gihou*, 32(2000)76 (in Japanese).
7. Imai A., *et al.*, Nippon steel technical report, 84(2001)1.
8. Takami A., SAE Paper 950746.
9. Hori, H., SAE Paper 972850.
10. Takahashi N., *Catalysts Today*, 27(1996)63.
11. Hachisuka I., SAE Paper 20011196.
12. Noda A., JSAE paper 20014525 (in Japanese).
13. Wiehl J. and Vogt C.D., *MTZ*, 64(2003)113.
14. Knon H., Brensheidt T. and Florchinger P., *MTZ*, 9(2001)662.
15. Rhodia, Homepage, http://www.rhodia.ext.imaginet.fr, (2003).
16. Eastwood P., *Critical topics in exhaust gas aftertreatment.*, Hertfordshire, Research Studies Press Ltd., (2000)33.
17. Tanaka T., *JSAE 20034493* (in Japanse).
18. Johnson Matthey, Homepage, http://www.jmcsd.com,(2003).
19. PSA Peugeot Citroen, Homepage, http://www.psa-peugeot-citroen.com (2003).
20. Tanaka T., 22nd International Vienna Motor Symposium, (2001)216.

The turbocharger and the exhaust manifold

11.1   Functions of the turbocharger

Internal combustion engines ignite air and fuel to produce energy that is converted to power. The waste created by the combustion is expelled. Compressors in the charging systems increase output by compressing the air used for combustion. There are three basic types of compressors, exhaust gas turbochargers, mechanically driven superchargers and pressure wave superchargers.[1] The latter two compress air using power supplied by the crankshaft, while the turbocharger is powered by the exhaust gas.

A turbocharger (Fig. 11.1) gives a small engine the same horsepower as a much larger engine and makes larger engines more powerful, increasing power output by as much as 40%.[2] Turbocharging was rapidly adopted for

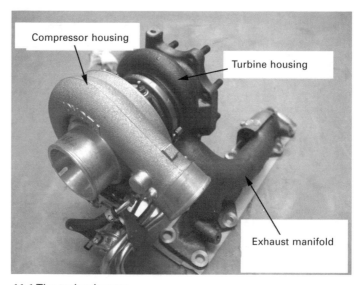

11.1 The turbocharger.

commercial diesel applications after the first oil crisis in 1973.[3] Stringent emission regulations mean that today, virtually every truck engine is turbocharged.

Turbocharged petrol engines for cars came into fashion because of their power, but their role in reducing emissions is now recognized. The introduction of a turbocharged diesel car in 1978 was the breakthrough for turbocharging in engines. Subsequent improvements in diesel engines for cars have increased efficiency, improved drivability to match that of petrol engines and reduced emissions.

The turbocharger is basically an air pump. It makes the air/fuel mixture more combustible by introducing more air into the engine's chamber which, in turn, creates more power and torque. It accomplishes this task by condensing or compressing the air molecules, increasing the density of the air drawn in by the engine.

Hot exhaust gases leaving the engine are routed directly to the turbine wheel to make it rotate. The turbine wheel drives the compressor wheel via the shaft. The typical turbocharger rotates at speeds of 200,000 rpm or more. The rotation of the compressor wheel pulls in ambient air and compresses it before pumping it into the engine's chambers. The compressed air leaving the compressor wheel housing is very hot, as a result of both compression and friction. The charge-air cooler reduces the temperature of the compressed air so that it is denser when it enters the chamber. It also helps to keep the temperature down in the combustion chamber.

The most recent turbochargers adjust the cross-section at the inlet of the turbine wheel in order to optimize turbine power according to load, a system known as variable geometry. The advantages of the turbocharger include a high power-to-weight ratio, so engines are more compact and lighter, a high torque at low engine speeds, which results in quieter engines, and superior performance at high altitudes. Currently, the primary reason for turbocharging is the use of exhaust gas energy to reduce fuel consumption and emissions.

11.2   The turbine wheel

11.2.1 Turbine and compressor designs

Figure 11.2 shows a cutaway of a turbocharger. Turbochargers consist of an exhaust gas-driven turbine and a radial air compressor mounted at opposite ends of a common shaft (Fig. 11.3) and enclosed in cast housings. The shaft itself is enclosed and supported by the center housing, to which the compressor and turbine housings are attached. The turbine section is composed of a cast turbine wheel, a wheel heat shroud and a turbine housing, with the inlet on the outer surface of the turbine housing. It functions as a centripetal, radial- or mixed-inflow device in which exhaust gas flows inward, past the wheel

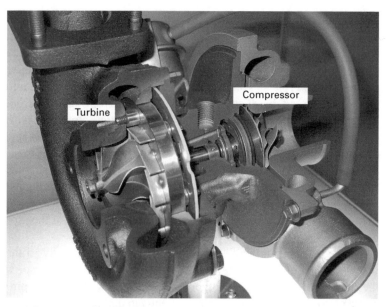

11.2 Cut away of turbocharger.

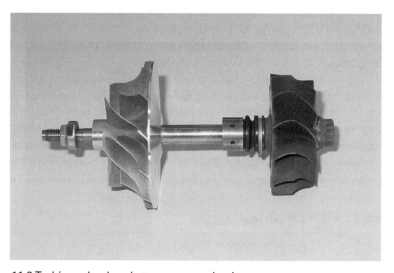

11.3 Turbine wheel and compressor wheel.

blades, and exits at the center of the housing. The expanded engine exhaust gas is directed through the exhaust manifold into the turbine housing. The exhaust gas pressure and the heat energy extracted from the gas cause the turbine wheel to rotate, which drives the compressor wheel.

The Ni-based super alloy Inconel 713C (see Table 11.1) is widely used for the turbine wheel.[4] A typical microstructure is shown in Fig. 11.4. GMR235,

Table 11.1 Chemical composition of materials (%) used in exhaust devices

	wt%	Ni	Fe	C	Cr	Mo	Co	W	Ta	Nb	Si	Al	V	Ti	Hf	Cu	Others
Turbine wheel	Inconel 713C	Balance	2.5	0.2	13.0	4.5	–	–	–	2.0	–	6.1	–	0.7	–	–	B0.01, Zr0.1
	GMR235	Balance	10.0	–	15.5	5.3	–	–	–	–	–	3.0	–	2.0	–	–	Mn0.1, B0.1
	Mar-M247	Balance	–	0.2	8.3	0.7	10.0	10.0	3.0	–	–	5.5	–	1.0	1.5	–	B0.015, Zr0.07
	Intermetallic compound TiAl	–	–	–	1.0	–	–	–	–	4.8	0.2	33.5	–	Balance	–	–	–
Compressor wheel	Al Alloy C355	–	0.2	–	–	–	–	–	–	–	5.0	Balance	–	0.2	–	1.5	Mn0.1, Mg0.5
	Ti-6Al-4V alloy	–	–	–	–	–	–	–	–	–	–	6.0	4.0	Balance	–	–	–
Turbine housing & exhaust manifold	Hi-Si nodular cast iron	–	Balance	3.8	–	–	–	–	–	–	3.0	–	–	–	–	–	–
	Si-Mo nodular cast iron	–	Balance	3.8	–	0.5	–	–	–	–	3.0	–	–	–	–	–	–
	Niresist nodular cast iron	30.0	Balance	–	5.0	–	–	–	–	–	5.0	–	–	–	–	–	–
	Ferritic cast steel	0.8	Balance	0.4	18.5	–	–	1.0	–	0.7	–	–	–	–	–	–	–
	Austenitic cast steel	12.0	Balance	0.3	20.0	–	–	1.0	–	0.7	0.2	–	–	–	–	–	–
Exhaust manifold	JIS-SUH409L	<0.6	Balance	<0.08	11.0	–	–	–	–	–	<1.0	–	–	~0.75	–	–	Mn<1.0
	JIS-SUS430J1L	–	Balance	<0.025	18.0	–	–	–	–	0.4	<1.0	–	–	Ti+Nb+Zr~0.8	–	0.4	Mn<1.0
	JIS-SUS444	–	Balance	<0.025	18.0	2.0	–	–	–	–	<1.0	–	–	Ti+Nb+Zr~0.8	–	–	Mn<1.0

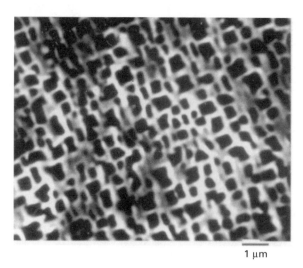

1 µm

11.4 The microstructure of Inconel 713C.

which reduces costs by increasing the iron content, is also used. For much higher temperatures, Mar-M247 is used. The response and combustion efficiency of the wheel in acceleration is related to the inertial moment, a function of the weight. The lower the weight, the lower the inertial moment and therefore the faster the response. Ceramic wheels[5] have been developed, but low toughness means that the blades must be thick, making it less easy to adjust the weight. A wheel made from the intermetallic compound TiAl by investment casting has been marketed.[6] It has a specific gravity of 3.9 g/cm^3, which is much lower than ordinary titanium alloy, and a tensile strength as high as 600 MPa at 700 °C.

The compressor section is composed of a cast compressor wheel, a backplate and a compressor housing, with the inlet at the center of the compressor housing. It is a centrifugal or radial-outflow device, in that the air flows outward, past the wheel blades, and exits at the outer edge of the housing. The rotating compressor wheel draws ambient air through the engine's air filtration system. The blades accelerate and expel the air into the compressor housing, where it is compressed and directed to the engine intake manifold through ducting. The compressor wheel does not have such a high heat resistance requirement, so a cast aluminum wheel (C355) is widely used. A cast Ti-6Al-4V alloy is also used for heavy-duty commercial diesels.[3]

11.2.2 Investment casting

Steel parts are more difficult to cast than cast iron parts because of shrinkage and gas porosity. The turbine and compressor wheels have very complicated shapes and high dimensional accuracy is important. Investment casting, often called lost wax casting, is therefore used to make the turbine wheel and the aluminum compressor wheel. A schematic illustration is shown in Fig. 11.5. The process involves precision casting to fabricate near-net-shaped metal parts from almost any alloy.

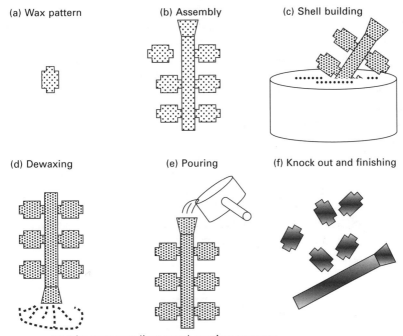

(a) Wax pattern

(b) Assembly

(c) Shell building

(d) Dewaxing

(e) Pouring

(f) Knock out and finishing

11.5 Investment (lost wax) casting process.

Investment casting is one of the oldest manufacturing processes. The Egyptians used it in the time of the Pharaohs to make gold jewelry (hence the name investment) some 5,000 years ago. The most common use in recent history has been the production of components requiring complex, often thin-walled castings. It can be used to make parts that cannot be produced by normal manufacturing techniques, such as complex turbine blades that are hard to machine, or aircraft parts that have to withstand high temperatures.

The process begins with the fabrication of a sacrificial pattern with the same basic geometric shape as the finished part (Fig. 11.5(a)). Patterns are normally made of investment casting wax, which is injected into a metal die. The wax patterns, normally more than 20 pieces, are assembled using the gate and runner system (b). The entire wax assembly is dipped in refractory ceramic slurry, which coats the wax and forms a skin (c). The skin is dried and dipping in the slurry and drying is repeated until a sufficient thickness is achieved. Once the refractory slurry has dried completely to become a ceramic shell, the assembly is placed in a steam autoclave to remove most of the wax (d). Just before pouring, the mold is pre-heated to about 1,000 °C to harden the binder. Pre-heating also ensures complete mold filling. Pouring can be done under gravity, pressure or vacuum conditions (e). When the metal has cooled and solidified, the ceramic shell is broken off by vibration or high-pressure water blasting (f). Next, the gates and runners are cut off, and sand blasting and machining finish the casting.

Investment castings often do not require any further machining because of the close tolerances. Normal minimum wall thicknesses are 1 to 0.5 mm for alloys that can be cast easily. Since the turbine and compressor wheels have a complex shape and the super alloy used for the turbine wheel is very hard, investment casting is the only process suitable for mass-production.

Ti alloys are more reactive with air at high temperatures than steels or Ni alloys. They also react with the crucible during melting, which results in contamination of the molten metal and therefore affects the properties of the castings. The TiAl turbine wheel is produced by a method that ensures that molten Ti does not come into contact with the crucible. Figure 11.6 schematically illustrates the process, which is known as Levicasting.[7] In this process, a magnetic force generated by an induction coil causes the molten TiAl to float, so that it does not touch the water-cooled Cu crucible, thus avoiding contamination during melting. The molten TiAl is dosed into the

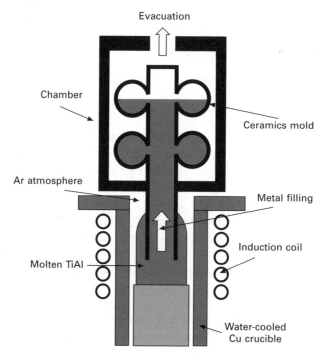

11.6 Ti casting using Levicast process.

bottom of the ceramic mold by pressurized Ar gas, and evacuation of the mold facilitates filling.

11.3   The turbine housing

11.3.1 Cast iron

The turbine housing must be resistant to oxidation and possess thermal fatigue resistance at high temperatures. Diesel engines have low exhaust gas temperatures and therefore use high Si nodular cast iron or Niresist cast iron. These housings are made by sand casting using sand cores.

The Si content of high-Si nodular graphite cast iron is about 14%, which raises the eutectoid transformation temperature (723 °C, the transformation temperature from austenite to pearlite). Transformation during operation causes thermal fatigue because of the transformation strain. By raising the transformation temperature, the added Si prevents thermal fatigue in the operating temperature range. The Si also forms a thick oxide scale on the surface which prevents oxidation corrosion at high temperatures. This alloy is also used in the exhaust manifold.

Austenitic nodular graphite cast iron is also used in the housing, and adding about 20% Ni makes the matrix of cast iron austenitic in a wide temperature range. Generally, this form is referred to as Niresist. Figure 11.7 shows the microstructure of Niresist. Round graphite particles can be seen in the austenitic matrix. The austenite structure has a high thermal expansion coefficient, and the lack of transformation under operating conditions gives superior resistance to thermal fatigue. Niresist is stronger than high-Si nodular graphite cast iron. The high strength and corrosion resistance mean that this alloy can be used in the exhaust manifold, which operates under red heat conditions. The high hardness and high thermal expansion coefficient are also suitable for reinforcing the piston-ring groove of the aluminum piston (see Chapter 3).

The thermal inertia of the turbine housing affects start-up emissions. Low thermal inertia means that the temperature of the catalytic converter rises quickly, so that it begins to convert pollutants early. This operational design requires greater heat resistance and thin walls in the turbo housings.

11.3.2 Cast steel

Petrol engines have much higher exhaust gas temperatures and use ferritic cast steels or austenitic cast steels for the turbine housing. Austenitic steel is stronger than ferritic steel, but the higher thermal expansion coefficient is a disadvantage. An austenitic steel with a lower thermal expansion coefficient has been developed.[8] Thin-walled, defect-free cast steel housings are preferable,

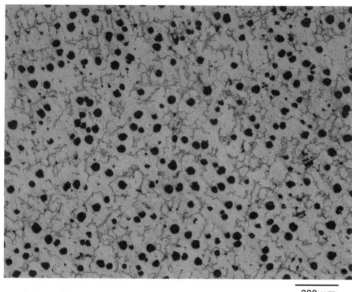

200 μm

11.7 Microstructure of Niresist nodular cast iron. The spheroidized graphite disperses in the austenite matrix containing a small amount of chromium carbide.

and are made by countergravity low-pressure casting,[9,10] which is an investment casting method that uses a vacuum to suck the molten metal through the ceramic mold.

11.4   The exhaust manifold

The exhaust manifold collects the exhaust gas and expels it through the exhaust pipe. Figure 11.8 shows the properties required. At present, the exhaust manifold must be capable of withstanding continuous operating temperatures as high as 900 °C. However, environmental and economic requirements will result in higher exhaust gas temperatures, so the thermal reliability of the exhaust manifold must be improved further. Traditionally, full load air/fuel conditions have been operating in the region of lambda = 0.9 for maximum engine power output and to maintain engine durability. Under these circumstances, excess fuel cools the engine, keeping the exhaust gas temperature below 1,000 °C. Moves towards operating conditions where lambda = 1 will eliminate this fuel cooling effect and exhaust temperatures will go up to 1,050 °C.

The location and light-up time of the catalyst are important factors given the increasingly stringent regulations on startup emissions. To activate the catalyst during startup, the exhaust gas temperature needs to be kept high

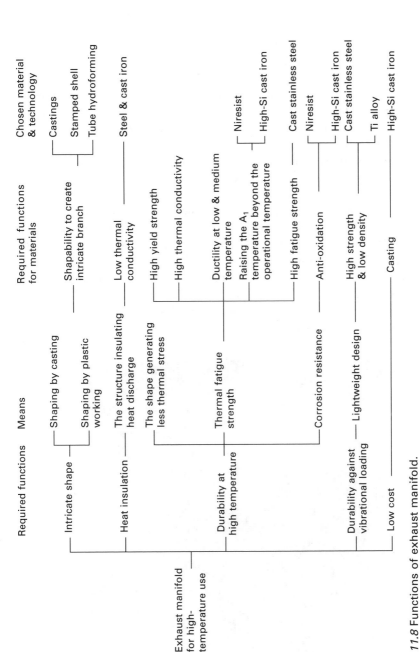

11.8 Functions of exhaust manifold.

until it reaches the catalytic converter, requiring the exhaust manifold to have thermal insulating properties. Temperature distribution in the manifold is complicated by exhaust gas recirculation and the installation of sensors, which result in large cold operating areas, as well as the air injection required for hydrocarbon combustion in the catalytic converter.

Exhaust manifold materials must have good fatigue strength under repeated thermal stress and be resistant to corrosion. Thermal stress causes plastic deformation and cracking takes place at low and intermediate temperatures, so the yield strength and ductility should be raised to restrict fatigue failure. Oxidation corrosion reduces the wall thickness of the manifold and the oxide residue that separates damages the turbine wheel and catalyst. Inhomogeneous corrosion also initiates fatigue cracks. Corrosion resistance is particularly important in diesel engine manifolds, because of the continuous flow of highly oxidizing gas generated by lean combustion. Vibrational loading is inevitable in the reciprocating engine. The heavy turbocharger and additional exhaust devices attached to the manifold increase the load, so high fatigue strength under vibrational loading is also necessary.

The intricate shape of the manifold can be made easily from cast iron (Fig. 11.9). Figure 11.10 shows a thin-walled manifold. High-Si ferritic nodular cast iron is used for operating temperatures up to 800 °C. Added Mo raises heat resistance and strength. Increasing V in the cast iron (Fe-3.3%C-

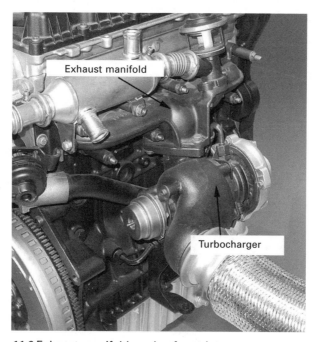

11.9 Exhaust manifold made of cast iron.

11.10 Thin walled exhaust manifold made of cast iron. The right half shows cutaway view.

4.2Si-0.5V-0.5Mo-3Mn)[11] is another method of improving intermediate temperature strength. This iron has a higher thermal conductivity and lower thermal expansion coefficient than Niresist cast iron. For much higher operating temperatures, up to 1,000 °C, Niresist nodular cast iron is used.

These castings are widely used and inexpensive. However, two other manifold technologies have evolved to solve the problems of weight and emission requirements. One is a fabricated stainless steel manifold.[12,13] This manifold is fabricated from stamped shells with welded or bent tube runners. The latter, shaped by hydroforming, are shown in Fig. 11.11. A double-walled air gap design is frequently used.[14] This protects against heat loss, reduces noise and is light. Austenitic steel has higher strength than ferritic steel, but despite its lower strength, ferritic steel is more widespread, typically JIS-SUH409L and SUS430J1L, because its low thermal expansion coefficient prevents the oxide scale from peeling off during repeated thermal cycles. Type 429Nb and SUS444 are used for higher temperature conditions.[15] In the double-walled manifold, it is common to use austenitic steel for the inner tube and ferritic steel for the outer tube.

The alternative to the fabricated stainless steel manifold is a cast steel manifold. Both ferritic and austenitic alloys, as listed in Table 11.1, are used.

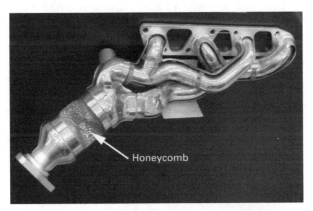

11.11 Fabricated exhaust manifold installing close-coupled catalyst produced by sheet metal forming. The outer shell is removed to show the double walled structure.

To reduce the weight, thin wall casting technologies such as countergravity and low pressure casting have been developed.[9,16,17] Currently, cast iron and cast steel are the material of choice for 20% of exhaust manifolds, with the remainder being fabricated steel manifolds.

11.5   Conclusions

Environmental considerations mean that exhaust gas must be clean and energy efficiency must be high. Exhaust gas temperatures are tending to increase in both diesel and petrol engines as a result of energy efficiency measures. The exhaust manifold and muffler work as an electronically controlled system, and exhaust gas temperatures above 950 °C subject the materials used to severe conditions. Current efforts are aimed at raising performance and durability without increasing costs, and reducing development lead time.

11.6   References and notes

1. *Automotive Handbook*, ed. by Bauer H., Warrendale, SAE Society of Automotive Engineers, (2000)440.
2. Honeywell, Homepage, http://www.egarrett.com, (2003).
3. BorgWarner Turbo Systems, Homepage, http://www.3k-warner.de, (2003).
4. Noda S., *Materia Japan*, 42(2003)271 (in Japanese).
5. Matoba K., *et al.*, SAE Paper 880702.
6. Kanai T., *et al.*, *Materia Japan*, 39(2000)193 (in Japanese).
7. Daido casting, Homepage, http://www.d-cast.jp, (2003).
8. Uyeda S., *et al.*, *Denkiseikou*, 72(2002)93 (in Japanese).
9. Yonekura T., Nagashima T. and Yamazaki H., *Sokeizai*, 37(1996)22 (in Japanese).
10. Hitchiner catalogue, (2003).
11. Suzuki N., *et al.*, *JSAE* 20034500 (in Japanese).
12. Inoue Y. and Kikuchi M., *Shinnitesu Gihou*, 378(2003)55 (in Japanese).
13. Miyazaki A., Hirasawa J. and Sato S., *Kawasakiseitetsu Gihou*, 32(2000)32 (in Japanese).
14. This design has been used in motorcycles for years.
15. Kikuchi M., *Tokushukou*, 49(2000)10 (in Japanese).
16. Takahashi N., *Kinzoku*, 62(1992)27 (in Japanese).
17. Itou K. and Otsuka K., *Jact news*, 9(2000)19 (in Japanese).

Age hardening A phenomenon in which rapidly cooled aluminum alloy and steel are hardened by ageing.

Ageing A phenomenon in which, after rapid cooling or cold working, the properties of metal (for example, hardness) are changed with the lapse of time.

Alloy A metallic substance that is composed of two or more elements.

Annealing Heat treatment consisting of heating and soaking at a suitable temperature followed by cooling under conditions such that, after return to ambient temperature, the metal will be in a structural state closer to that equilibrium.

Austempering Heat treatment involving austenitizing followed by step quenching, at a rate fast enough to avoid the formation of ferrite or pearlite, to a temperature above Ms and soaking to ensure partial or entire transformation of the austenite to bainite. The purpose of the operation is to prevent the generation of a strain and quenching cracks and to give strength and robustness.

Austenite Solid solution of one or more elements in gamma iron.

Bainite An austenitic transformation product found in some steels and cast irons. It forms at temperatures between those at which pearlite and martensite transformations occur. The microstructure consists of ferrite and a fine dispersion of cementite.

Carbide A compound of carbon and one or more metal elements. Carbides having not less than two alloying elements as necessary components is called double carbide.

Carbon potential A term indicating the carburizing capacity of an atmosphere for heating steel. This is expressed by the carbon concentration of a steel surface when it is in equilibrium with the gas atmosphere at the temperature.

Cementite Iron carbide with the formula Fe_3C.

Crystal structure For crystalline materials, the manner in which atoms or ions are arrayed in space. It is defined in terms of unit cell geometry and the atom positions within the unit cell.

Decarburization   Depletion of carbon from the surface layer of a ferrous product.

Diffusion   Mass transport by atomic motion.

Eutectic   A change from one solution to a structure in which not less than two solid phases are closely mixed in the process of cooling, or the structure generated as a result of the reaction. When the concentration of alloying metal element is smaller than that in the eutectic solution, it is called hypo-eutectic, and when larger, hyper-eutectic.

Eutectoid   A transformation from one solid solution to the structure in which not less than two solid phases are closely mixed in the process of cooling or the structure generated as a result of the transformation. When the concentration of alloying metal element is smaller than that in the eutectoid solution, it is called hypo-eutectoid, and when larger, hyper-eutectoid.

Fatigue   Failure, at relatively low stress levels, of structures that are subjected to fluctuating and cyclic stresses. The fatigue life is defined as the total number of stress cycles that will cause a fatigue failure at some specified stress amplitude.

Ferrite   Solid solution of one or more elements in alpha iron or delta iron.

γ iron   Stable state of pure iron between A_3 (911 °C) and A_4 (1,392 °C). The crystal structure is face-centered cubic. It is paramagnetic.

Grain boundary   The interface separating two adjoining grains having different crystallographic orientations.

Grain size   Characteristic size of a grain revealed in a metallographic section. Generally it is expressed with the grain size number obtained by either the comparison method or the cutting method.

Hardness   The measure of a material's resistance to deformation by surface indentation.

Hydrogen embrittlement   A phenomenon in which the toughness of steel is deteriorated by the absorbed hydrogen. This is usually generated when pickling, electroplating or the like is carried out, and further, with the existence of a tensile stress, often results in cracking.

Impact energy   A measure of the energy absorbed during the fracture of a specimen subjected to impact loading. Charpy and Izod impact tests are used.

Intergranular fracture   Fracture of polycrystalline materials by crack propagation along the grain boundaries.

Intermetallic compound   A compound of two metals that has a distinct chemical formula. On a phase diagram it appears as an intermediate phase that exists over a very narrow range of compositions.

Low-temperature annealing   An annealing carried out at a temperature not higher than the transformation temperature for the purpose of lowering the residual stress or softening. This is sometimes carried out at a temperature not higher than recrystallization temperature.

Malleability   This term is used when plastic deformation occurs as the result of applying a compressive load. The plastic deformation under a tensile load is referred to as ductility.

Patenting   An isothermal heat treatment applied to medium- and high-carbon steel wire prior to its final drawing operation. This process generates steel wire having high tensile strength. This produces strong wire such as piano wire. Patenting consists of passing the wire through tubes in a furnace at about 970 °C. After austenitizing the wire at 970 °C and the rapid cooling to 550–600 °C generate a very fine pearlitic microstructure. The resulting ferrite with a fine distribution of carbide has a very high ductility and can be cold drawn with total reduction in diameter of 90%. The cold drawn wire may achieve tensile strength levels in excess of 1.6 GPa without becoming brittle.

Phase   A homogeneous portion of a system that has uniform physical and chemical characteristics.

Precipitation   A phenomenon in which a crystal of a different phase is separated from a solid solution and grows.

Quenching   Operation which consists of cooling a ferrous product more rapidly than in still air. The use of the term specifying the cooling conditions is recommended, for example air-blast quenching, water quenching, step quenching, etc.

Residual stress   A stress that exists inside metal though no external force or thermal gradient is acting. When a heat treatment is carried out, thermal stress or transformation stress due to the difference of cooling rate is produced inside and outside of the material and these combined remain inside the material as stress. The residual stress is also produced by cold working, welding, forging, etc.

Segregation   A phenomenon in which alloying elements or impurities are unevenly distributed, or its state.

Single crystal   A crystalline solid for which the periodic and repeated atomic pattern extends throughout the crystal. A single crystal does not include grain boundaries.

Solid solution   Homogeneous, solid, crystalline phase formed by two or more elements.

Solution treatment   Heat treatment intended to dissolve previously precipitated constituents and retain them in the solid solution.

Spheroidal (spheroidized) carbide or globular carbide   Carbides in a globular form.

Spheroidizing   Geometric development of the carbide particles, such as the cementite platelets, toward a stable spherical form.

Strain ageing   An ageing occurring in cold worked materials.

Sub-zero treatment or deep freezing   Heat treatment carried out to transform the retained austenite into martensite after quenching, and consisting of cooling to and soaking at a temperature below ambient.

Superalloy Alloys capable of service at high temperatures, usually above 1,000 °C. Ni and Co alloys are normally included.

Supercooling (undercooling) An operation in which metals are cooled down to the transformation temperature or the solubility line or lower so that transformation and precipitation may be entirely or partly prevented.

Temper embrittlement (brittleness) Brittleness which appears in a certain quenched and tempered steel, after soaking at a certain temperature or during slow cooling through these temperatures. The primary temper embrittlement produced by tempering at about 500 °C and the secondary temper embrittlement produced by slow cooling after tempering at even higher temperatures are called high-temperature temper embrittlement. The temper embrittlement in the case of tempering at temperatures around 300°C is called low-temperature temper embrittlement.

Tempering Heat treatment applied to a ferrous product, generally after quench hardening, or another heat treatment to bring the properties to the required level, and consisting of heating to specific temperatures ($<Ac_1$) and soaking one or more times, followed by cooling at an appropriate rate.

Toughening An operation in which steel is turned into troostite or sorbite structure by tempering at a comparatively high temperature (about 400 °C or higher) after quench hardening. This increases the ratio of the elastic limit to the ultimate tensile strength (yield ratio). This is also referred to as thermal refining.

Transformation A crystal structure is changed into another crystal structure by the rise or fall of temperature. Temperature at which a change of phase occurs and, by extension, at which the transformation begins and ends when the transformation occurs over a range of temperatures.

Transgranular fracture Fracture of polycrystalline materials by crack propagation through the grains.

T6 The temper designation system is used for all forms of wrought and cast aluminum and aluminum alloys except ingot. For heat-treatable alloys the following designations are used. T1: cooled from an elevated-temperature shaping process and naturally aged to a substantially stable condition. T2: cooled from an elevated-temperature shaping process, cold worked and naturally aged to a substantially stable condition. T3: solution heat-treated, cold worked and naturally aged to a substantially stable condition. T4: solution heat-treated and naturally aged to a substantially stable condition. T5: T1 + artificial age. T6: solution heat-treated and artificially aged (T4 + artificial age). T7: solution heat-treated and overaged/stabilized. T8: T3 + artificial age. T9: T6 + artificial age. T10: T2 + artificial age.

Appendix A: international standards conversion table for alloys

Table A.1 compares JIS and other alloy standards. Blanks indicate no direct comparison, although similarities in chemical composition among materials may be identified.

Table A.1 Comparison of Japanese industrial standard (JIS) with other standards

Japan	USA			UK	Germany	France	ISO
JIS	AA	AISI/SAE	ASTM	BS	DIN	NF	ISO
AC4B	333.0	331	333.0	LM-13		A-S12UNG	
AC8A	336.0	321	336.0	LM-28		A-S18UNG	
AC9B				LM2			
ADC12	383.0				G(GK)-AlSi9Cu3		
FC200			Class No. 30	200.0		FLG200	
FC350			Class No. 50	350.0		FLG350	
FCD700-2			100-70-03	350.0	GGG-70	FGS700-2	700-2
S45C		1045		C45	C45	C45	C45
S50C		1049		C50	C50	C50	C50
S55C		1055		C55	C55	C55	C55
SCM420				708M20			
SCM435		4137			34CrMo4	34CrMo4	34CrMo4
SUS304		304	S30400	304S31	X5CrNi1810	Z7CN18.09	11.0
SKD11			D2	BD2		X160CrMoV12	
SKD61			H13	BH1B	X40CrMoV51	X40CrMoV5	40CrMoV5
SUH1			S65007	401S45		Z45CS9	X45CrSi93
SUH3						Z40CSD10	
SUH35				349S52		Z53CMN21.09AZ	X53CrMnNiN219
SUJ2		52100	52100		100Cr6	100Cr6	B1 or 100Cr6

Appendix B: function analysis table

The engine parts explained in this book have various functions, and the function analysis tables used in several chapters examine the function of a part and the associated requirements for materials and manufacture. For example, the camshaft has to drive accurately to open and close the valves while rotating at high velocity. Figure B.1 analyses the three fundamental functions required, which include: (i) the camshaft should drive the valve accurately even at high rotational velocities; (ii) the camshaft itself should rotate at high speed without torsion and bending; (iii) camshaft manufacture requires precision at low cost.

The third column lists the means for meeting the requirements of each function, and the fourth column lists the properties required of the materials used. The shape of the camshaft, which is another aspect that must be taken into consideration in the design and manufacture of camshafts, is not included in this table.

The fifth column lists the materials and material technologies suitable for meeting the functional and property requirements of camshafts. These are known as technological seeds. For instance, steel is preferred over aluminum for the shaft portion because of its higher rigidity. Various methods can be used to harden the cam lobe, such as quench-hardening of forged steel or cast iron, carburizing of forged steel, chilled cast iron, remelting of the cam lobe portion of the cast iron camshaft or sintering. Quench-hardening may be used for forged steel or cast iron, but remelting is applicable only to cast iron, so remelting cannot be used if the designer plans to use a lightweight steel. Generally, a part performs several functions simultaneously, and the mechanical designer must choose the most suitable material and technology on the basis of analysis and experience.

Reference

1. Wright I.C., *Design Methods in Engineering and Product Design*, Berkshire, McGraw-Hill Publishing Company, (1998) 221.

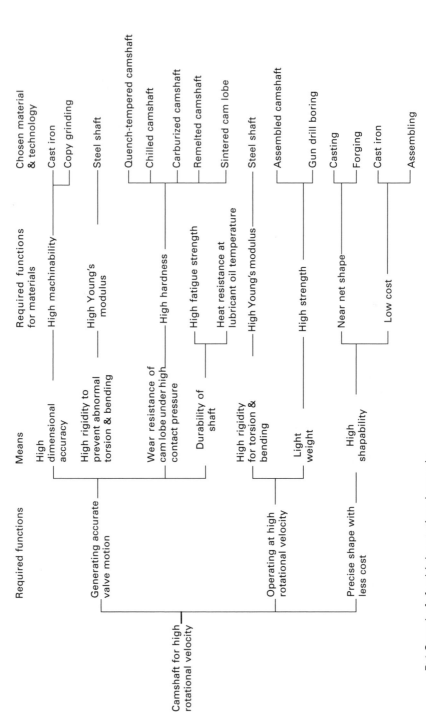

B.1 Camshaft for high rotational speeds.

Appendix C: the phase diagram

Equilibrium phase diagram: the crystal structure of a metal often changes with temperature. When a pure metal absorbs a certain amount of another element, it becomes an alloy and the crystal structure will change. The phase diagram is a map that shows the variations in crystal structure across a wide temperature range.

Figure C.1 is the binary phase diagram of the alloy consisting of iron and

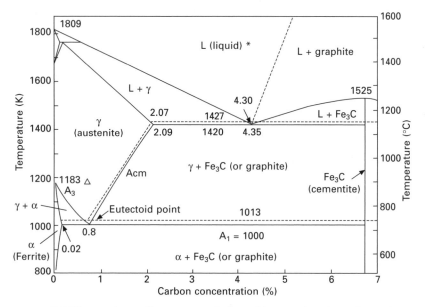

C.1 Binary phase diagram consisting of iron and carbon. A steel containing 0.8% carbon transforms from austenite into a mixture of ferrite and cementite. This is called eutectoid transformation. The 0.8% carbon steel is especially called eutectoid steel. The temperature at which the eutectoid transformation takes place is termed the eutectoid point.

The annealed eutectoid steel consists only of pearlite. The steels having a higher carbon content above the eutectoid composition are called hyper-eutectoid steels. The hyper-eutectoid steels comprise both cementite and pearlite. The steels having a lower carbon content below the eutectoid composition are called hypo-eutectoid steels. The hypo-eutectoid steels comprise both ferrite and pearlite. We can roughly judge the carbon content in a steel through observing its microstructure.

Transformation temperatures change with carbon content. Each boundary line at which the crystal structure changes has a particular name. A_1: The horizontal line at 723 °C (1000 K). Acm: the oblique line between γ and $\gamma + Fe_3$ C. A_3: the oblique line between γ and $\gamma + \alpha$. Also, the transformation temperature shifts a little either in cooling or in heating. To distinguish it, a suffix c is attached in heating, while r in cooling. These are indicated such as Ar_1 or Ac_1.

carbon. The carbon content is shown on the horizontal axis and temperature on the vertical axis. Pure iron is represented on the left (carbon content = 0%), and carbon content increases to a maximum of 7% on the right-hand side of the diagram.

The phase diagram indicates the equilibrium states at various compositions and temperatures, and is also referred to as an equilibrium phase diagram. In thermo-dynamics, the state of equilibrium is reached when there is no net heat exchange between an object and its surroundings. For instance, when a glass of water at 10 °C is placed in a room at 30 °C, the temperature of the water will rise until it is the same as that of the room. This is the equilibrium state, and it will remain stable unless the temperature of the room is changed.

The phase diagram displays equilibrium states on a temperature vs. composition plane. For instance, in Fig. C.1, an iron containing 4.3% carbon is liquid at 1,450 °C (indicated by *). Below 1,154 °C, it is solid for all compositions. Crystal structures change in the solid state. The boundary lines in the phase diagram separate the different crystal structures. The area enclosed by a boundary line has the same crystal structure throughout.

Crystal structures given by equilibrium transformations: Table C.1 summarizes the characteristics of the typical crystal structures shown in Fig. C.1. Ferrite (Fig. C.2 (a)) exists in the narrow portion on the left side in Fig. C.1. Ferrite (α-iron) has a bcc structure where iron atoms (white circles) are arranged as shown schematically in Fig. C.2 (a).

Table C.1 Characteristics of typical crystal structures[1].

Name	Microstructure	Characteristics
Austenite	γ iron	γ solid solution containing a carbon content below 2.06%. This transforms to pearlite below 723 °C. Alloys having this structure are tough, corrosion resistive and paramagnetic.
Ferrite	α iron	α solid solution dissolving small amounts of carbon (0.02% at 723 °C, and 0.006% at room temperature). This phase is soft, ductile and ferromagnetic.
Cementite	Iron carbide (Fe_3C)	Hard and brittle iron-compound containing 6.67%C. This phase is ferromagnetic at room temperature, while ferrimagnetic above A_0 transformation (215 °C).
Pearlite	Eutectoid precipitates of α and carbide	A lamellar structure formed through A_1 transformation, comprising ferrite and cementite.

There is a vertical line at 6.67%C, which represents the composition of a carbide called cementite. Since the ratio Fe_3C of iron to carbon does not

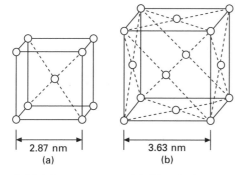

C.2 (a) Bcc structure of ferrite. (b) Fcc structure of austenite. Austenite (γ-iron) has a fcc structure. The interaction between atoms determines metal structures. A metal includes countless crystal lattices comprising such atomic arrangements. One lattice has a size of 3–4 nm. The difference in the crystal structure corresponds to the difference in the atomic arrangement.

change up to the melting temperature, it is shown by a straight line. Cementite is very hard. It raises hardness and strength when dispersed finely in the iron matrix.

The state γ + Fe$_3$C, where austenite and cementite coexist, is stable below 1,147 °C. The mixed state α + Fe$_3$C, consisting of ferrite and cementite, appears below 723 °C. A steel of 0.8% carbon is austenite at 900 °C. It changes to a mixed state comprising ferrite and cementite below 723 °C. This mixed state is called pearlite.

Changes in crystal structure are referred to as transformation. The transformation of 0.8% carbon steel from austenite to a mixture of ferrite and cementite is referred to as the eutectoid transformation, and 0.8% carbon steel is frequently called eutectoid steel. The temperature at which the eutectoid steel transforms is termed the eutectoid point. Steels with a carbon content above eutectoid steel are called hyper-eutectoid steels, whereas steels with a lower carbon content are called hypo-eutectoid steels.

From the phase diagram, it can be seen that pure iron transforms from ferrite to austenite at 910 °C (allotropic transformation). Figure C.3 shows the microstructures of irons of various compositions, obtained by etching polished iron alloys with acids and viewed under 100 times magnification.

Figure C.3 (a) is a typical ferrite of a 0.01% carbon steel. Only linear grain boundaries are observable (see Appendix G). Each grain boundary separates single crystals. Figure C.3 (b) shows the microstructure of a 0.35% carbon steel, comprising ferrite and pearlite. Pearlite is a mixture of ferrite and cementite. Pearlite displays a lamellar microstructure similar to a herringbone pattern under microscopy.

Figure C.3 (c) shows the microstructure of a 0.8% carbon steel consisting of pearlite. In the region of the mixture of α + γ, the amount of cementite

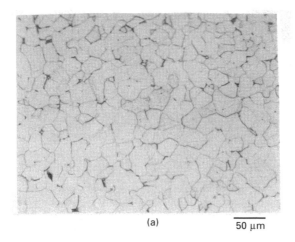

(a)   50 μm

C.3 (a) Microstructure of a 0.01% carbon steel. Grain boundaries are observable.

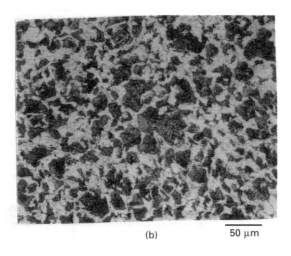

(b)   50 μm

C.3 (b) Microstructure of a 0.35% carbon steel (hypo-eutectoid steel). The white portions are ferrite. Pearlite is gray because it is fine.

increases with increasing carbon content. The amount of ferrite inversely decreases.

Coarse grain size in steel lowers impact strength considerably, so it is very important to measure and control the grain size. The grain size is adjusted by heating the steel in the austenite temperature region. However, since austenite transforms to ferrite and cementite below 723 °C (Fig. C.1), the original austenite grain boundary is not observable at room temperature and a different technique is needed to see the austenite grain boundary at high temperatures.

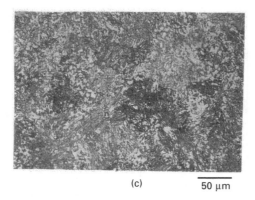

(c)   50 μm

C.3 (c) Microstructure of a 0.8% carbon steel (eutectoid steel). Pearlite showing a herringbone appearance containing white ferrite and gray cementite.

The specimen is quenched from the austenite state and etched by a picric reagent. Figure C.3 (d) indicates the austenite grains thus revealed.

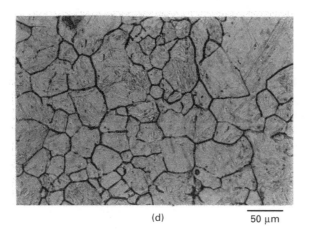

(d)   50 μm

C.3 (d) Austenite grains.

Cast iron: Figure C.3 (e) is a micrograph of 3% C cast iron. (Cast iron refers to an iron containing carbon content above 2%.) Crystallized graphite flakes are observable. When the carbon content is as high as this, the carbon (see Appendix D) appears as graphite. The specimen is not etched, so the microstructure of the matrix is not observable.

Martensite: under slow cooling, austenite transforms into ferrite at 910 °C as shown in the phase diagram. The crystal structure changes from that of Fig. C.2 (b) to that of Fig. C.2 (a). The atoms rearrange into a dissimilar

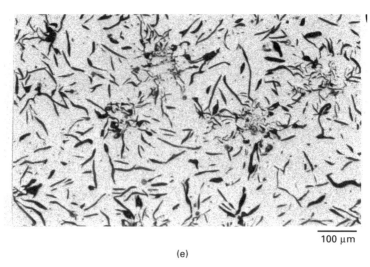

100 μm

(e)

C.3 (e) Flake graphite distribution of a 3% C gray cast iron. The detailed microstructure of the matrix is not observable because this is unetched.

configuration through atomic diffusion. However, if austenite is cooled rapidly down to room temperature, the rearrangement of atoms is restricted. In this case, the transformation illustrated in the phase diagram does not take place.

Slow cooling to temperatures below 723 °C transforms austenite into a mixture of ferrite and cementite as indicated in Fig. C.1. However, rapid quenching using water or oil transforms austenite into a crystal structure called martensite, and the transformation shown in the phase diagram does not take place. Figure C.3 (f) indicates a micrograph of martensite, showing a needle-like shape. Martensite is not shown in the phase diagram because it does not appear as an equilibrium phase, it is a crystal structure of a nonequilibrium state. Martensite is hard and greatly strengthens steels. This heat treatment is known as quench-hardening (see Appendix F).

Metastable equilibrium phase diagram: the iron-carbon system phase diagram is sometimes called a double phase diagram because it indicates two transformations in one diagram. One is the equilibrium iron-graphite system and the other is the metastable iron-cementite system. In the equilibrium state, carbon exists in iron as graphite. However, the iron combines with carbon to form cementite during cooling after solidification. It is predicted from the iron-graphite system that cementite decomposes into the equilibrium state of iron and graphite during prolonged heating. However, this takes place very sparingly, and so the cementite phase can exist as an almost stable phase, known as a metastable phase.

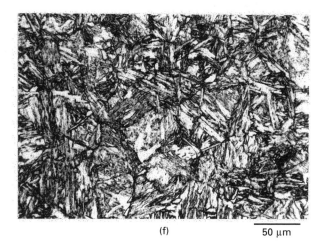

(f)   50 μm

C.3 (f) Martensite.

Since graphite is difficult to nucleate in iron, most iron-carbon alloys are in the metastable state. Therefore, in the iron-carbon phase diagram, the solid line indicates the iron-cementite system while the dashed line represents the iron-graphite system. When cooling is rapid, cast iron solidifies according to the metastable iron-cementite system. The carbon crystallizes as cementite (see Chapter 4), and the resultant hard microstructure is called chill. By contrast, when cooling is slow, the carbon crystallizes as graphite in accordance with the iron-graphite system, resulting in gray cast iron (see Appendix D).

References

1. Ochiai Y., *Sousetsu kikaizairyou*, ver. 3, Tokyo, Rikougakusha Publishing, (1993) (in Japanese).
2. Kinzokubinran, ver. 5, *Nippon Kinzoku Gakkai*, Tokyo, Maruzen Co. Ltd., (1990) 509 (in Japanese).

Appendix D: types of cast iron

Iron alloys (see Fig. C.3 (e)) containing 2% carbon and above are called cast iron. Cast irons show microstructures with dispersed graphite (crystal carbon) in the matrix. The graphite distribution depends on the solidification rate in the mold. The microstructure of the matrix is influenced by the cooling rate after solidification.

The natural differences among cast irons are due to dispersed graphite in the microstructure. A high carbon content, over 3%, lowers the melting temperature of iron, so that it makes casting easy. Fine patterns in the mold can be transferred precisely into the cast part because cast irons expand when graphite crystallizes. During machining, graphite works as a chip-breaker to give high machinability and dimensional accuracy. Since graphite itself works as a solid lubricant during machining, the cutting tool is unlikely to seize, another factor in the high machinability. Dispersed graphite also gives cast irons a high damping capacity.

Casting methods control the graphite shapes. Various graphite shapes are shown in Fig. D.1. We can change the shape from flaky (Fig. C.3 (e) or type I in Fig. D.1) to spherical or nodular graphite (type VI in Figure D.1). Since the dispersed graphite flakes work as micro-notches, flaky graphite cast irons are brittle. By contrast, nodular graphite cast irons have a high ductility due to the spherical graphite shape (see Chapter 2).

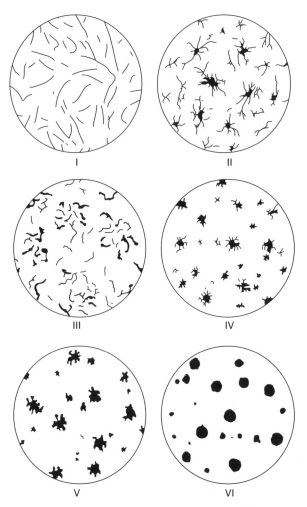

D.1 Various graphite shapes in cast irons[1].

The matrix of cast iron has the chemical composition of high-carbon steels. Additional heat treatment can change the microstructure of the matrix without changing the graphite shape and distribution. Figure D.2 shows an as-cast state without additional heat treatments.

50 µm

D.2 Microstructure of a flaky graphite cast iron. The herringbone pattern of the matrix is pearlite (a lamellar structure of cementite and ferrite). When Fig. C.3 (e) is etched, this microstructure is observable. The spacing between the cementite and ferrite is comparatively wide. Except for the graphite, the microstructure of the matrix is similar to that of a 0.8% C eutectoid steel in appearance.

Silica (SiO_2) sands are generally used as a mold material for sand casting. Various binders to fix the sand particles have been developed. The binder material controls not only the strength of the mold during casting but also mold decay after casting.

Unlike aluminum alloys (see Appendix J), permanent mold casting for cast iron is rare, because metal molds have a short lifetime under the high melting temperature. The rapid cooling rate of metal molds also causes chill in the casting (see Chapter 5). Table D.1 lists examples of cast irons for various engine parts.

High-silicon nodular cast iron: increasing Si up to about 14% raises the eutectoid transformation temperature (723 °C; the transformation temperature from austenite to pearlite). During operation of a part, if the temperature rises and falls around the transformation temperature, transformation and inverse transformation are repeated, which can result in thermal fatigue failure. Adding Si raises the transformation temperature to prevent thermal fatigue. It also forms a thick oxide scale that prevents oxidation corrosion at high temperatures. This alloy is therefore used in exhaust manifolds (see Chapter 11).

Table D.1 Cast irons for engine parts

Types	Parts
Gray cast iron	Cylinder block, camshaft
Ferritic gray cast iron	Housing cover
Nodular cast iron	Steering knuckle, crankshaft
High-Si nodular cast iron	Exhaust manifold
Austenitic nodular cast iron	Turbine housing, exhaust manifold
Vermicular cast iron	Exhaust manifold, cylinder block, piston ring
Malleable cast iron	Camshaft, connecting rod, transmission gear
Alloy cast iron	Camshaft
	Tappet
	Cylinder liner
	Valve rocker arm
	Cylinder head
	Cylinder block

Austenitic nodular cast iron: Adding Ni up to about 30% makes the matrix austenite over a wide temperature range, and this cast iron is referred to as Niresist. Thermal fatigue resistance is superior to that of high-silicon nodular cast irons, so Niresist is used for turbocharger housings, which operate under red heat conditions. It is also used for ring inserts in aluminum pistons, because the high hardness with the high thermal expansion coefficient reinforces the piston-ring groove (see Chapter 3). The austenite structure has a high thermal expansion coefficient (see also Chapter 11).

Vermicular graphite cast iron: the graphite shape is short (type III in Fig. D.1). It has intermediate properties between those of flaky graphite cast iron and nodular cast iron (see also Chapter 2).

High alloy cast iron: Fe, C and Si are basic elements in gray cast irons. Cast irons with additional alloying elements are called high alloy cast irons (see Chapter 3), and elements such as Cu, Cr, Mo, Ni, Sn, P, etc., are added to improve the properties of the matrix.

Malleable cast iron: this cast iron first solidifies as chilled cast iron. Subsequent high-temperature annealing graphitizes the cementite in the chill to give malleability. Since the graphite becomes nearly spherical in shape, ductility increases. Chilled cast iron is also called white cast iron, after the color of the fracture surfaces of the cast parts.

References

1. *JIS Handbook of Iron and Steel*, Tokyo, The Japanese Standards Organization, (1996) (in Japanese).
2. Zairyouno Chishiki, *Toyota Gijutsukai*, (1984) 46 (in Japanese).

Appendix E: steel-making and types of steel

Automotive engines use various kinds of iron, mostly iron alloys called steels. Steels generally contain carbon and other alloying elements. Figure E.1 illustrates the process for producing irons and steels from iron ore. The ore (mainly iron oxide) is chemically reduced to molten iron in a blast furnace at high temperatures in a CO atmosphere. The resultant iron is called pig iron, and still contains carbon (about 4%) and other impurities. Pig iron has a chemical composition close to cast iron and is brittle due to the high level of impurities. A converter or an electric furnace removes the impurities to give the required composition of steel. This part of the process is called steel-making. Waste steel scraps are put directly into the steel-making process, by-passing the blast furnace.

The steels produced are cast into various ingot shapes (billet, bloom or slab), mainly using the strand, or continuous, casting process, which solidifies the melt continuously. The steel ingots are converted into the desired shape by rolling. Even after refining, non-metallic inclusions such as Al_2O_3, MnS, (Mn, Fe)O $\cdot$ SiO_2, remain in the steel. These inclusions are internal defects and are sometimes the cause of cracking. To obtain high quality steels, molten steel must be further refined by a secondary refining process, as described in Chapter 9. Steels are classified in various ways, as listed in Table E.1. Steels can fall into more than one category, depending on the classification.

The mechanical properties of steel are very sensitive to carbon content, and carbon steels are classified into low-, medium- or high-carbon types. Of all the different steels, those produced in the greatest quantities fall into the low-carbon classification. These generally contain less than about 0.3% C and are unresponsive to heat treatment, so are strengthened by cold working.

Medium-carbon steels have carbon concentrations in the range of about 0.3 to 0.5%. Austenitizing, quenching and tempering improve the mechanical properties. High-carbon steels normally have carbon contents in the range of 0.50 to 1.5%. These are the hardest and strongest, but least ductile, steels. Carbon steels that contain only residual concentrations of impurities other than carbon and a little manganese are called plain carbon steels. For alloy steels, more alloying elements are added.

Table E.2 classifies steels according to chemical composition. The typical alloying elements along with examples of applications are also listed. Plain carbon steels can lower the material cost of a part. Generally, the higher the amounts of alloying elements, the higher the price of the steel. However, engine parts often use high alloy steels because of the high strength requirements.

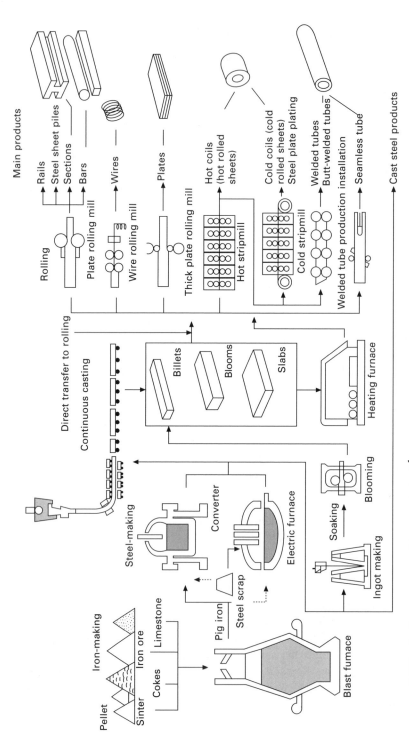

E.1 Production process of iron and steel.[1]

Table E.1 Classifications of steels[1]

Classification	Specification
Strength	High-tensile-strength steel, ultra-high-tensile-strength steel, high-yield-strength steel.
Shape	Steel sheet, thick plate, tube, bar, wire, foil.
Characteristic	Heat-resisting steel, wear-resisting steel, free-cutting steel, stainless steel, case-hardening steel, steels for general engineering purposes.
Usage	Structural steel, tool steel, high-speed steel, music wire, bearing steel, spring steel, railway steel, steel tubes for piping.
Composition	Ultra-low-carbon steel, low-carbon steel, medium-carbon steel, high-carbon steel, low-alloy steel, high-alloy steel, Si-Mn steel, Ni steel, Cr steel, Cr-Mo steel.
Manufacturing process	Hot-finished steel, cold-finished steel, cast steel, forged steel, wrought steel, hot-rolled steel, cold-drawn steel, hard-drawn wire steel.
Heat treatment	Normalized steel, maraging steel, thermally refined steel, non-thermally refined steel, hardenable alloy steel.
Microstructure	Ferritic steel, ferrite-pearlite steel, austenitic steel, bainitic steel, martensitic steel, dual-phase steel.
Surface modification	Coated steel, tinplate, blackplate, hot-dip zinc-coated steel sheet.
Steel-making process	Electric furnace alloy steel, acid Bessemer carbon steel, basic open-hearth carbon steel.
Deoxidizing	Killed steel, rimmed steel, Al killed steel, Ti killed steel.

References

1. *Tetsugadekirumade*, Nippon Tekkou Renmei, (1984) (in Japanese).

Table E.2 Types of steels classified by chemical compositions[1]

	Specification	Composition characteristics	Products
Carbon steel	Ultra-low-carbon steel	C < 0.12%	Stamped body for car, refrigerator, and washing machine. Electrical wire. Tin-coated plate. Zinc-coated plate.
	Low-carbon steel	0.12–0.3%	Bars. Shaped steels and sheets for ships, building, cars, and bridges. Tubes for gas or water. Nails. Thread.
	Medium-carbon steel	0.3–0.5%	Wheels for train. Shaft. Gear. Spring.
	High-carbon steel	0.5–0.9%	Rail. Wire-rope. Spring.
	Tool steel	0.6–1.5%	Shaver. Cutting tool. File, Chip. Spring. Drill.
Low-alloy steel	Si steel	Si; 0.5–5%	Electric motor. Transformer.
	Structural alloy steel	Ni; 0.4–3.5%. Cr; 0.4–3.5%. Mo; 0.15–0.7 %.	Bolt. Nut. Shaft. Gear.
	Alloy tool steel	Cr < 1.5% W < 5%. Ni < 2%.	Cutting tool. Die. Punch. File. Chisel. Belt saw.
	Bearing steel	Cr; 0.9–1.6%	Bearing.
	High-tensile-strength steel	Cu, Ni, Cr < 1%	Building. Bridge. Ship. Railway. Mining.
High-alloy steel	Stainless steel	Ni; 8–16% Cr; 11–20%	Furniture. Cutlery. Chemical plants.
	Heat-resisting steel	Ni; 13–22% Cr; 8–26%	Engine. Turbine.
	High-speed steel	W; 6–22%. V, Co	Drill. Cutting tool.

Appendix F: creating various properties through heat treatment

Heat treatment procedures transform the crystal structure and therefore alter the microstructure through heating or cooling (see Appendix C). Figure F.1 illustrates several heat-treatment procedures. The vertical axis indicates the treatment temperature and the horizontal axis the treatment time. Various microstructures can be obtained by changing the heating temperature and the cooling speed after heating. The table lists the names of heat treatments, purposes, resulting microstructures and typical parts. Figures F.2 (a) to (e)

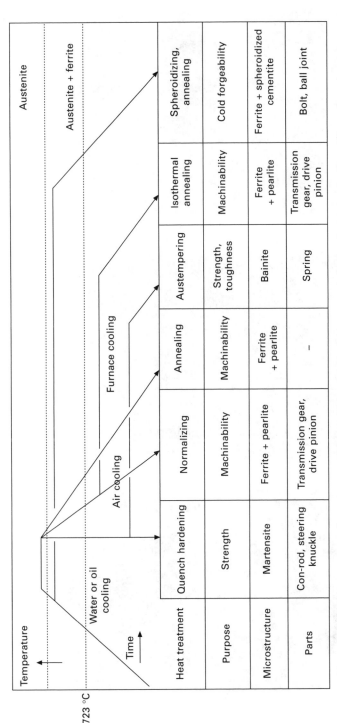

Heat treatment	Quench hardening	Normalizing	Annealing	Austempering	Isothermal annealing	Spheroidizing, annealing
Purpose	Strength	Machinability	Machinability	Strength, toughness	Machinability	Cold forgeability
Microstructure	Martensite	Ferrite + pearlite	Ferrite + pearlite	Bainite	Ferrite + pearlite	Ferrite + spheroidized cementite
Parts	Con-rod, steering knuckle	Transmission gear, drive pinion	–	Spring	Transmission gear, drive pinion	Bolt, ball joint

F.1 Heat treatment diagram.[1] The vertical axis indicates temperature and the horizontal axis time. The microstructure becomes austenite + ferrite above the temperature of the lower broken line (723 °C) and becomes austenite above the upper broken line. The detailed transformation temperatures shown as the upper line differ with carbon concentration, while the lower line is always located at 723 °C. The two lines coincide at the composition of the eutectoid steel. The upper line moves to the high-temperature side with decreasing carbon content in the hypo-eutectoid steel range.

(a)

50 μm

(b)

50 μm

(c)

50 μm

(d)   50 μm

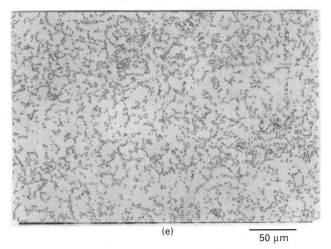

(e)   50 μm

F.2(Left and above) (a) Martensite of a medium carbon steel JIS-S45C (Fe-0.45%C-0.25Si-0.8Mn) having a hardness of 661 HV. HV indicates Vickers hardness number. (b) Sorbite of JIS-S45C having a hardness of 256 HV. (c) Normalized S45C having a hardness of 186 HV. (d) Annealed S45C having a hardness of 161 HV. (e) S45C after spheroidizing annealing having a hardness of 213 HV. Courtesy of Sanyo Special Steel Co., Ltd.

show the typical microstructures of a carbon steel, JIS-S45C, after various heat treatments.

Quench hardening: when heating medium or high carbon steel at the austenite temperature (around 900 °C) followed by quenching in water or oil, a hard microstructure called martensite appears (see Appendix C). This is because the rapid quenching suppresses the eutectoid transformation that gives rise to the pearlite microstructure. This heat treatment is called quench hardening. Figure F.2 (a) shows a martensite microstructure.

The chemical composition of steel is important if the appropriate hardness value cannot be obtained by quenching. The addition of elements such as Mn, Cr, and/or Mo gives a hard martensitic structure even if the cooling rate is slow. These alloying elements therefore give high strength at slow cooling rates, for instance, in thick parts that cannot be quenched rapidly, even under water-cooling. In high-alloy steels, such as high-Cr die-steels, air-cooling after heating can produce very high hardness because the martensitic transformation can take place without quenching.

When austenite transforms to martensite, the crystal expands in accordance with the structural change. For example, the percentage volume change is about 4.5% in S45C steel. The expansion depends on the carbon content, being predicted by the following empirical equation: $4.75 - 0.53 \times C\%$. This volume change causes distortion in the quenched part. If the distortion is not relaxed, cracking sometimes occurs, and this is referred to as quenching crack.

Tempering: steels are brittle just after quench hardening, so quenched steel is usually tempered to give the appropriate ductility without lowering the strength and hardness. Tempering destroys some of the hard martensite microstructure to give a sorbite microstructure. The tempering temperature is in the range of 150 to 700 °C. Quench hardening and tempering are generally carried out as a series of heat-treatment procedures. Figure F.2 (b) shows a sorbite microstructure.

Steels acquire various material properties through changing the combination of tempering temperature and time. Figure F.3 shows the change in mechanical

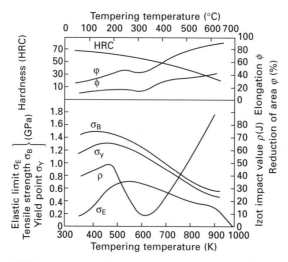

F.3 Master tempering curve of a 0.3% C steel.[2]

properties of a 0.3% C steel against tempering temperature under a constant tempering time of one hour. This is the master tempering curve, and similar curves are produced for various kinds of steels so that the conditions for adjusting mechanical properties can be found at a glance.

Tempering increases ductility. However, the impact value becomes very low at a tempering temperature of around 300 °C (Fig. F.3), which is called temper embrittlement. This temperature should not be chosen except in special cases, for instance, to improve the spring property of high Si steels (see Chapter 7). Tempering at higher temperatures, around 600 °C, is used to homogenize the microstructure while avoiding temper embrittlement, and is called high-temperature tempering. After high-temperature tempering, the tempered steel components must be cooled rapidly because they pass through 300 °C. Carburized and quenched steel parts are tempered at around 150 °C to obtain ductility with sufficient hardness, because tempering at around 600 °C eliminates the intended high hardness.

Retained austenite: even after quenching, a small amount of metastable austenite remains. Figure F.4 shows a microstructure where the austenite remains in martensite. This retained austenite is unstable at room temperature. Under high applied stress, the retained austenite transforms to martensite, causing distortion. This transformation under load is unfavorable in parts where high dimensional accuracy is necessary, such as bearings.

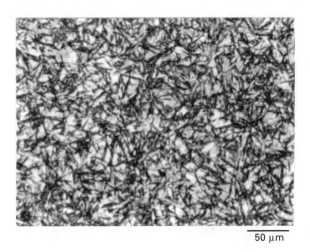

50 μm

F.4 Leaf-like martensite in the retained austenite (white portions) of quenched JIS-S45C.

The retained austenite transforms to martensite if cooled below room temperature. Sub-zero treatment, when a quenched steel part is immersed in liquid nitrogen, removes the retained austenite. Shot peening is an alternative method of causing martensitic transformation. The temperature at which martensitic transformation takes place decreases with increasing carbon content, and falls below room temperature in steels with a composition above 0.6% C. Therefore, a high amount of retained austenite is likely to appear in high-carbon steels and carburized steels. Although retained austenite is unstable and causes distortion when martensitic transformation takes place during operation, it is reported that a small amount of retained austenite increases toughness considerably.

Normalizing: austenite steel transforms into a mixture of ferrite and pearlite during slow cooling. If the cooling rate is slightly faster than air-cooling, the transformed pearlite becomes fine. It raises the strength, but the value is lower than that produced by quench-hardening. This treatment is called normalizing. Figure F.2 (c) shows a normalized microstructure.

Annealing: when steels are gradually cooled from the austenite state by turning off the power to the furnace, the resultant pearlite becomes coarse and soft. This heat treatment is called annealing. Figure F.2 (d) shows an annealed microstructure.

Isothermal annealing: austenitizing steels first and then cooling and holding at just below the A_1 temperature generates a ferrite and cementite microstructure. The original austenitizing produces a soft annealed state within a shorter period of time than annealing at a temperature below A_1.

Austempering: holding a part at a constant temperature during cooling below A_1 gives rise to bainite, a tough microstructure with an intermediate hardness between pearlite and martensite. This treatment is referred to as austempering and is often used in spring steels for cushion springs, etc.

Spheroidizing annealing: this treatment spheroidizes cementite to increase ductility and malleability. It is used in a cold forging billet and is appropriate for severe working conditions. Figure F.2 (e) shows a microstructure after spheroidizing annealing. The spheroidized cementite of a bearing steel (JIS-SUJ2) is shown in Chapter 9 at high magnification. Various methods have been proposed for spheroidizing.

References

1. *Zairyouno Chisiki, Toyota Gijutsukai*, (1984) 46 (in Japanese).
2. *Kinzoku Binran,* ver. 5, *Nippon Kinzoku Gakkai*, Tokyo, Maruzen Co. Ltd., (1990) 549 (in Japanese).

Appendix G: mechanisms for strengthening metals

The plastic deformation of metals (see Appendix K) is caused by moving dislocations. The dislocation is a line defect around which atomic misalignment exists in the crystal lattice. If the atomic arrangement of a crystal is regarded as layered planes, dislocations move along the atomic plane. When dislocations move, the crystal lattice (Fig. C.2) slips and displaces as shown in Fig. G.1. The atomic planes on which dislocations move are called slip planes.

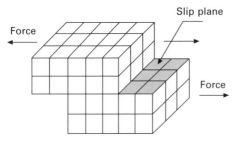

G.1 Slip in crystal lattices.[1]

Using a thin metal film transparent to an electron beam, dislocations are observable as black strings under electron microscopy. Figure G.2 is a photograph showing dislocations. The distorted atomic arrangement of a dislocation scatters the transmitted electron beam, which causes the linear shadow in the photograph. The misalignment of the atoms continues like a

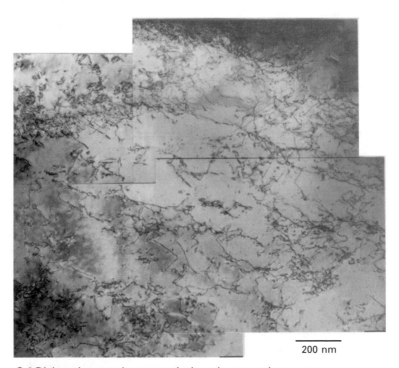

G.2 Dislocations under transmission electron microscopy.

string in a crystal lattice, so it is called a dislocation line. Various hardening mechanisms for metals have been examined microscopically, including solution hardening, precipitation hardening, dislocation hardening, grain size reduction hardening and microstructural hardening.

Figure G.3 represents dislocations as cars. If the cars (dislocations) run smoothly, the situation corresponds to the high deformability of a soft metal. Here, the slip plane of pure iron can be likened to a well-paved road covered by asphalt, upon which the cars flow smoothly.

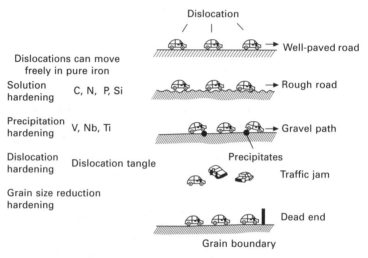

G.3 Strengthening methods of metals.[2]

Hardening

Solution hardening: in an alloy, the solid solution state has solute atoms randomly dispersed among the solvent atoms. The solute atom has a different atomic radius from the solvent iron atoms and therefore strains the crystal lattice plane. This situation is shown in Fig. G.4. Using the car analogy, this situation is likened to a rough road, where the cars cannot run smoothly. This situation, where dislocations cannot move easily, corresponds to low deformability. Iron in this state is hard and strong, and this mechanism is referred to as solution hardening. The higher the content of added elements (C, N, P and/or Si), the harder the iron alloy becomes, for instance 0.3% carbon steel is stronger than a 0.1% carbon steel.

Precipitation hardening: if elements such as V, Nb and/or Ti are added, they combine with the dissolved carbon or nitrogen to precipitate the associated carbide or nitride compounds. Since these precipitates introduce large internal strain, strength increases. Here we can compare the atomic planes to a gravel path with various sizes of stones. If the radius of the stones is at least a

(a)   (b)   (c)

G.4 Schematically illustrated lattice strain in solid solutions. The radius of the white circle (atom) is different from that of the black circle.[1]

quarter of the tire size and these stones are densely dispersed, the driver has to take a detour. Precipitates that increase strength range in size from around $10^{-3} \mu$m (a hundred atoms) to a few hundred μm. Age hardening (Chapter 3) of aluminum alloys results from the internal strain introduced by incoherent precipitation. This method of hardening was discovered accidentally by A. Wilm in 1906.[3]

Dislocation hardening: the number of dislocation lines increases with strain. This corresponds to an increase in car numbers. Increased traffic density causes traffic jams and accidents sometimes inhibit smooth traffic flow. This mechanism is called dislocation hardening or work hardening. Stamped sheet-metal parts obtain high strength during shaping as a result of this mechanism.

Grain size reduction hardening: a metal comprises a great number of crystals. One single crystal in a polycrystalline metal is referred to as a grain. The size of the grains ranges from a few to a few hundred μm (see Fig. C.3 (a)). At the boundary between the grains, the lattice planes of neighboring grains are not continuous with each other. This situation corresponds to a dead-end street. The car cannot pass through, or the dislocation cannot move through the grain boundary, and thus the metal is strengthened. The smaller the grain size, the greater the number of dead ends, hence the stronger the iron.

Microstructural hardening: microstructural hardening is caused by dispersed crystals (for example; martensite and bainite, etc.). A well-known example is a dual-phase steel sheet. It has a duplex microstructure containing both hard martensite and soft ferrite, showing adequate toughness and deformability. Equilibrium or non-equilibrium transformation can generate this hardening.

References

1. Ochiai Y., *Sousetsu Kikaizairyou*, ver. 3, Tokyo, Rikougakusha Publishing, (1993) (in Japanese).
2. Tanino M., *Feramu*, 1 (1996) 41 (in Japanese).
3. Wilm A., *Metallurgie*, 8(1911) 223.

Appendix H: surface modification

Surface modification or surface treatment changes the surface properties of a metal and is carried out for various purposes. The techniques used for engine parts are summarized in Table H.1. Most parts use some sort of surface modification. The treatment is sometimes carried out close to the completion stage, but as this generally raises the cost, designers often try to avoid it if possible. However, surface modifications sometimes give rare and desirable characteristics, since they are powerful means of improving material functions.

Reference

Nippon Piston Ring Co. Ltd., Catalog (in Japanese).

Table H.1 Surface modifications for engine parts

Name		Content	Characteristics	Hardness	Parts
Nitriding	Gas	Nitriding under NH$_3$	A harder surface than carburizing. High wear resistance.	1100–1200 HV	Piston ring, cylinder liner, tappet, rocker arm, valve.
	Salt bath	Nitriding in the molten salt.	A harder surface than carburizing. High wear resistance.	1100–1200 HV	
	Ion	A workpiece is charged under the mixed gas of nitrogen and hydrogen. The ionized nitrogen atom collides with the work surface to nitride the work.	Less distortion due to the low-temperature treatment. To adjust the compound and diffusion layers is easy.	1100–1200 HV	
Plating	Hard Cr	Electro-plating in a chromic acid solution.	Wear resistance, relatively cheap.	800–1000 HV	Piston ring, cylinder liner, aluminum cylinder block
	Soft metal	Plating in various metal electrolytes.	Solid lubrication property. Effective at running-in. Sn, Cu, and Ag.	–	Piston, plain bearing thrust washer
	Composite Ni	Composite Ni plating containing hard ceramics particles (Si$_3$N$_4$, SiC, WC, etc.). To increase the hardness, P is added in the electrolyte.	Wear resistance. Dispersed hard particles improve scuff resistance.	950–1000 HV	Piston ring, aluminum cylinder block
PVD		Evaporating a metal under vacuum. The evaporated metal sticks to the substrate. An ionized evaporated metal is also used through discharging.	A compound film having special property. Wear resistance. Scuff resistance. CrN, TiN, etc.	1800–2100 HV (In case of CrN)	Piston ring, rocker arm, valve lifter.

Table H.1 Continued

Name		Content	Characteristics	Hardness	Parts
Chemical conversion coating	Fe_3O_4	Immersing a steel workpiece into a hot alkali salt solution. Fe_3O_4 film is formed.	Effective at running-in	–	Piston ring
	Mn phosphating	Immersing a steel workpiece into hot manganese phosphate solution. A phosphate film is formed electrolessly.	Oil retention property due to the porous film. Effective at running-in.	–	Piston ring, cylinder liner, gear.
	Zn phosphating	Immersing a steel workpiece into hot zinc phosphate solution. A phosphate film is formed electrolessly.	Oil-retention property due to the porous film. Effective at running-in. Anti-rust.	–	Piston ring
	Chromic acid treatment	Immersing a workpiece in a chromic acid solution forms the film.	Corrosion resistance.	–	Camshaft cover, cylinder block (primer coating for painting).
Steam treatment		Heating and oxidizing a steel workpiece under saturated steam.	Decreasing friction. Wear resistance.	–	Valve seat, camshaft.
Anodizing		An aluminum workpiece is placd as the anode and electrolyzed in a sulfuric or phosphoric acid. A thin aluminum oxide film is formed.	Wear resistance. Corrosion resistance.	250–300 HV (In case of hard anodizing)	Piston, rocker arm.
Sulfurizing		FeS_2 coating in a salt bath	Initial wear, oil retention	–	Gear, shaft
Quenching	Flame	Local heating with oxygen-	Wear resistace. Less	600–700 HV (In	Tappet, camshaft

Name		Content	Characteristics	Hardness	Parts
		acetylene flame followed by quenching	decarburization and surface oxidation due to the short heating period.	case of hardenable cast iron)	
	Induction or laser heating	Local heating with high-frequency current followed by quenching.	Fatigue strength increase with the retained stress at the surface. Anti-pitting, anti-scuffing and wear resistance.	600–650 HV (JIS-S50C)	Camshart, crankshaft.
	Carburizing	Carbon atoms are doped into the surface of a low-carbon steel part under an atmosphere containing CO.	Tough core with a hard surface. Wear resistance. Anti-fatigue.	700–800 HV (JIS-SCM415)	Rocker arm, gear, camshaft, con-rod, crankshaft
Remelt chill		The surface of a gray cast iron part is remelted. It rapidly solidifies to cause chill.	Fine carbide. Wear resistance.	750–850 HV (Low alloy cast iron)	Camshaft, rocker arm, floating seal
Thermal spray	Gas	The spray metal is melted by oxygen-acetylene gas and is blown by compressed air.	Thermal spray of Mo, stainless steel, bronze, etc.	630–870 HV (Mo spray)	Piston ring, rotor housing, synchronizer ring, cylinder liner, shift fork.
	Plasma	Plasma arc melts a powder, then the melt is sprayed with inert gas to form a surface film.	Thermal spray of ceramics, cermet, super alloy, cemented carbide, etc.	700–760 HV (Mo+Ni base self melting alloy)	
	HVOF (high-velocity oxygen fuel)	A fine powder is melted by an oxygen mixed gas, then the melt is sprayed at a high velocity with a special gun to form a surface film.	Thermal spray of ceramics, cermet, super alloy, cemented carbide, etc.	600–750 HV (CrC/NiCr)	

Table H.1 Continued

Name	Content	Characteristics	Hardness	Parts
Resin coating	A polyamideimide or polybenzoimidasol resin is mixed with MoS_2 solid lubricant. The thinned resin with a solvent is sprayed. A hard surface film is formed through baking.	Inhibiting the aluminum adhesion at the top ring groove.	–	Piston ring, piston, bearing
Shot peening	Peening the surface with small steel shot.	Fatigure strength increases due to the compressive residual stress. This also improves the corrosion, wear, and pitting resistances.	–	Valve spring, gear

Appendix I joining technology

Figure I.1 classifies joining methods for metals. There are three main types of bonding: welding, mechanical bonding and adhesive bonding. Welding technologies are classified as fusion welding, pressure welding or brazing. Fusion welding joins two or more metal parts through melting and solidifying. The parts must be heated to melt them, but the welding is carried out without added pressure. Methods of fusion welding are classified according to heat source, such as gas welding, arc welding, laser beam welding, etc.

Pressure welding creates a join through exposing the bonding portion to pressure. It is performed either at room temperature or above the melting temperature. Ultrasonic welding, explosive welding and cold pressure welding are carried out with little or no heating. Diffusion bonding uses the property that clean surfaces spontaneously weld together on contact. Additional heating results in a stronger bond. Resistance welding uses an electric current targeted at the joining portion to melt it. Gas pressure welding uses oxygen and acetylene gas heating, and induction welding uses a high-frequency induction current. Friction welding uses the heat caused by the adiabatic shear of rubbing surfaces.

Brazing and soldering use filler metals that have a lower melting temperature than the parts to be joined, so the substrate parts do not melt. Capillarity helps the molten filler metal to penetrate into the narrow gap at the joint. Brazing is carried out at temperatures above 450 °C, while soldering is done below 450 °C, the difference being due to the melting temperature of the filler metal.

Reference

1. Matsumoto J., *Yousetsu Gakkaishi*, 63 (1994) 76 (in Japanese).

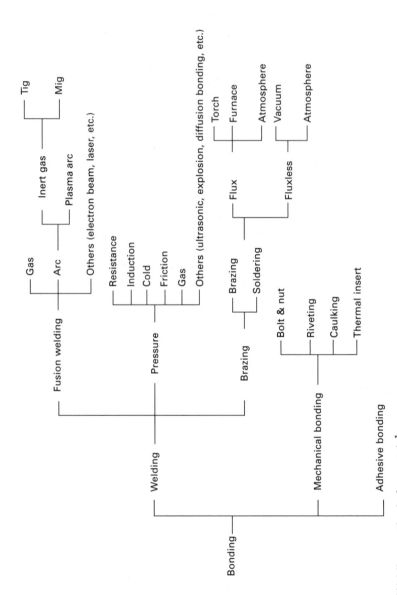

I.1 Welding methods for metals[1]

Appendix J: aluminum casting

Engine parts use various aluminum alloys. Most of them are cast parts. Figure J.1 lists various casting processes, classifying them according to mold type and the method used to apply pressure. Table J.1 summarizes the characteristics of the different casting technologies. In sand casting and gravity die casting, the weight of the molten metal itself fills the mold cavity.

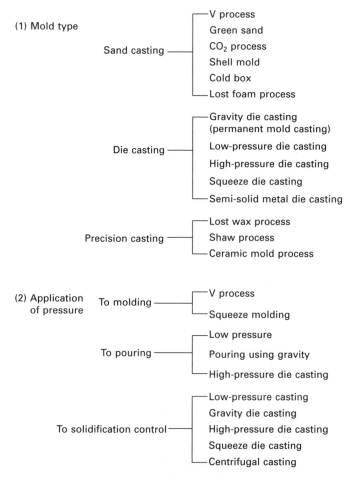

(1) Mold type

Sand casting
- V process
- Green sand
- CO_2 process
- Shell mold
- Cold box
- Lost foam process

Die casting
- Gravity die casting (permanent mold casting)
- Low-pressure die casting
- High-pressure die casting
- Squeeze die casting
- Semi-solid metal die casting

Precision casting
- Lost wax process
- Shaw process
- Ceramic mold process

(2) Application of pressure

To molding
- V process
- Squeeze molding

To pouring
- Low pressure
- Pouring using gravity
- High-pressure die casting

To solidification control
- Low-pressure casting
- Gravity die casting
- High-pressure die casting
- Squeeze die casting
- Centrifugal casting

J.1 Various casting process for metals.

Pressure die casting injects the pressurized melt into the die cavity. There are low-pressure and high-pressure die-casting techniques. High-pressure die casting injects the melt rapidly using hydraulic pressure, in either a cold-chamber or a hot-chamber process. In hot-chamber die casting, the cylinder and piston used to inject the molten metal into the die are immersed in the

Table J.1 Casting methods for aluminum alloys

	Sand casting	Gravity die casting (permanent mold casting)	Low-pressure die casting	High-pressure die casting			
				Conventional high-pressure diecasting	Vacuum die casting under the cavity pressure of 5 kPa	Squeeze die casting	Semi-solid metal die casting
Pressure (MPa)	Gravity	Gravity	20	100	100	70–150	100
Dimensional accuracy	Low	Medium	Medium	High	High	High	High
Minimum thickness (mm)	3	3	3	2	2	4	3
Quality							
Primary Si size in case of hyper-eutectic Al-Si (μm)	30–100	30–50	30–50	5–20	5–20	10–50	10–50
Gas content (cm^3/100 g)	0.2–0.6	0.2–0.6	0.2–0.6	10–40	1–3	0.2–0.6	0.2–0.6
Blow holes	Medium	Few	Few	A lot	Few	Few	Few
Shrinkage defects	Less than a few	Less than a few	Less than a few	A lot at thick portion	A lot	Few	Few
T6 treatment	Possible	Possible	Possible	Impossible	Possible	Possible	Possible
Welding	Possible	Possible	Possible	Impossible	Possible	Possible	Possible
Pressure tight	Low	Good	Good	Good after resin impregnation	Good	Excellent	Excellent
Productivity*	100	50	40	100	100	50	100
Life time of the mold*	Mold pattern has long life	150	150	100	100	70	100
Cost*	150	150	200	100	110	130–170	110

*The ratio where conventional high-pressure die casting is 100.

molten metal. This is used only for small thin castings in zinc and some magnesium alloys. By contrast, cold-chamber die casting is used for large parts made from aluminum, magnesium and brass. The aluminum cylinder block is made by this process. Unlike the hot-chamber machine, the metal injection system in this technique is in contact with the molten metal only for a short period of time.

The casting defect porosity is attributed to two main sources, solidification shrinkage and gas entrapment. Most alloys have a higher density in the solid state that in the liquid state. As a result, shrinkage porosity occurs during solidification. The shaped material resulting from conventional high-pressure die casting contains gas defects and pores.

Modified high-pressure die casting technologies have raised the quality of parts through reducing casting defects. One method is to control the atmosphere in the cavity. In PF (pore free) die casting, blowing oxygen to the molten aluminum eliminates hydrogen through the chemical reaction between oxygen and hydrogen. In vacuum die casting, the evacuation of the cavity prevents oxidation and enables a smooth metal flow. To keep the vacuum level in the cavity high, air leakage is prevented by sealing with a heat-resistant rubber. Figure J.2 is a schematic illustration of a vacuum die casting installation. The die is perfectly enclosed in the cover[2] for evacuation. The current process has enabled the production of a large weldable part for an automobile body, with a thickness of 2 mm and length of 2 m. Figure J.3[3] is a typical example of a part made by vacuum die casting.

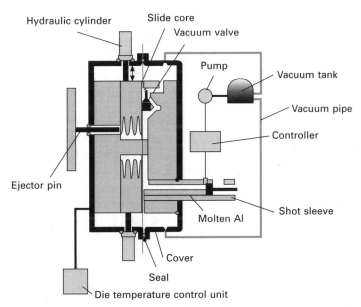

J.2 Schematic figure of an advanced high-pressure die casting.[2]

J.3 Thin-walled frame made by high-pressure vacuum die casting. The motorcycle with the frame installed is also shown on the right.

In squeeze die casting, the injected aluminum is squeezed in the mold just before solidification. Squeezing reduces the dissolved gas content and gives a high heat transfer from the melt to the mold, so that the melt is cooled rapidly, resulting in a fine microstructure. The quality obtained by this method approaches the level of forged parts.

Recently, a new set of casting technologies have been developed, called semi-solid-metal die casting, thixo casting or rheo casting. These are high-pressure die casting methods[4] where a half solidified melt is injection molded using a die-casting machine or a screw-driven injection machine. The common feature is that a shaped component is formed by manipulating specially prepared metallic alloys while they are roughly half solid and half liquid. The fine dendrite microstructure and low dissolved gas content give high strength as well as weldability and T6-treatability. Figure J.4 compares the quality and cost of several casting methods, together with forging. High quality means that the shaped material has a fine microstructure and few casting defects. The diagram gives a rough comparison.

The strength of aluminum alloys differs substantially depending on whether age hardening is applicable or not. During the age-hardening treatment, the

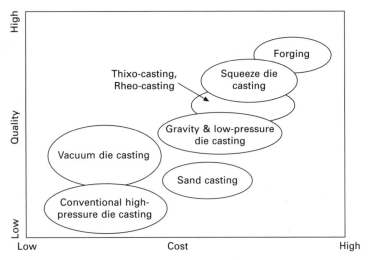

J.4 Quality and cost for several casting methods.[1]

alloy is usually heated to around 500 °C. When the dissolved gas content in a cast part is high, it forms blisters during heating. High gas content prevents the casting from age hardening. Welding is also difficult when the gas content is high. Low-temperature heating, below 250 °C, does not cause blisters even if the dissolved gas content is high. Table J.1 includes the gas contents given by the different casting methods. Figure J.5 illustrates the manufacturing steps for cast parts, from the planning stage to quality inspection stage after finishing. Casting can be used to mass-produce complex shapes.

References

1. *Aluminum Handbook*, ver. 5, Tokyo, Keikinzoku Kyoukai, (1994) 187 (in Japanese).
2. Kurita H., *et al.*, SAE paper 2004-01-1028.
3. Yamagata H., *Keikinzoku*, 53(2003)309 (in Japanese).
4. Vinarcik E.J., *High Integrity Die Casting*, New York, John Wiley & Sons, (2003)67.

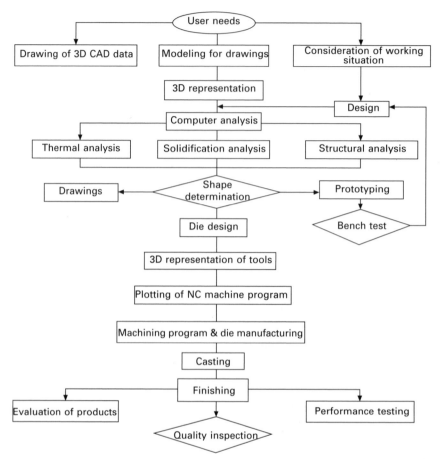

J.5 The process route to develop cast parts.[1]

Appendix K: elastic deformation and plastic deformation

Figure K.1 shows the stress-strain relation of a carbon steel. Metals generally follow Hooke's law at low stress or strain, and undergo elastic deformation. High stress or strain spreads or elongates the metal. Most metals can be shaped freely without failure. We call this property plasticity. Deformation to a high strain value above that of elastic deformation is called plastic deformation. Processing using plastic deformation is called plastic working.

The graph in Fig. K.1 is obtained by pulling a carbon steel wire on a tensile testing machine. Stress appears as a reaction force. OA is the range showing elastic deformation. Upon unloading, the stress applied within this range returns the wire to its original length. The point A is referred to as the yield point. The stress A is called the yield stress (indicated by as σ_y), and indicates where the elastic property of the test piece yielded to the pulling force. It is also known as the elastic limit, since elastic deformation is possible up to this limit. When the wire is strained beyond point A, the stress drops a little down to point B, and then increases towards point C. The stress decrease to B is due to the fact that the plastic deformation takes place faster than the pulling speed given by the testing machine. On further straining, a stress maximum appears at C, then the wire snaps. The stress C is referred to as the ultimate tensile strength (σ_{UTS}).

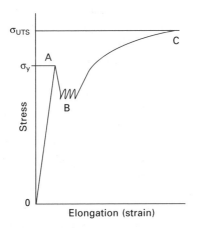

K.1 Stress-strain curve of a carbon steel during tensile testing. OA: elastic deformation.

Figure K.2 shows the stress to strain relation of pure aluminum. In this case, there is no clear yield point similar to A in Fig. K.1. The deformation gradually proceeds from an elastic to a plastic mode and the stress and strain relation seems to show a straight line near point a. Thus, it looks like an

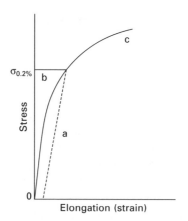

K.2 Stress-strain curve of pure aluminum. Elastic deformation ends and plastic deformation starts at somewhere between the point 0 and b.

elastic deformation. However, the elastic limit is not definite in this type of curve so, for convenience, we measure the stress at a small plastic strain and use the value as the yield stress. The stress at the plastic strain of 0.2% is frequently used. It is indicated as $\sigma_{0.2\%}$. This value is also referred to as the proportional limit. The σ_{UTS} is defined in the same manner as for Fig. K.1.

The stress-strain type shown in Fig. K.1 is observable in annealed or quench-tempered carbon steels, along with some non-ferrous alloys. However, most carbon steels and non-ferrous alloys show a curve like that in Fig. K.2. It is worth mentioning that the yield stresses listed in handbooks, etc., do not distinguish whether they refer to a clear elastic limit, like that in Fig. K.1, or to a proportional limit, like that in Fig. K.2.

When the elastic limit is used as the yield point, it is a well-defined point. However, when the yield point relates to the proportional limit, it must be recognized that a small plastic deformation has already taken place at the proportional limit $\sigma_{0.2\%}$. For example, in designing a bolt, if we regard the proportional limit value as the elastic limit value, the plastic deformation at $\sigma_{0.2\%}$ means that sufficient axial stress cannot be obtained, and therefore the design should not use an allowable stress around $\sigma_{0.2\%}$ for this bolt.

Metals inevitably contain dislocations, and when the metal experiences load, the dislocations move and breed even at a fairly low stress. This means that plastic deformation occurs at low stress, so most metals do not show a distinct elastic limit during straining and the curve represented in Fig. K.2 applies.

The definite elastic limit of a steel is observable when the dissolved carbon or nitrogen immobilizes the dislocations. A sufficiently high stress will cause the dislocations to move simultaneously, and this high stress corresponds to the elastic limit, as represented by the curve in Fig. K.1.

The small plastic deformation caused by unintentional dislocation motion is referred to as micro yielding (see Chapters 3 and 7). In the plastic working of a coil spring, a number of dislocations are introduced (work hardening, Appendix G). If the spring is used just after plastic working, it will sag and lose spring property because the dislocations move under loading. Low-temperature annealing prevents sagging and improves the spring property because the carbon or nitrogen atoms in the steel trap and immobilize the dislocations. High temperature annealing is not indicated for this type of spring because it decreases the dislocation density and thus lowers the spring property. This method of preventing micro yielding is called low-temperature anneal-hardening.

Micro-yielding is microscopic plastic deformation that takes place at stress levels below the macroscopic yield stress. Fatigue is closely related to micro-yielding. Repeated loading, even below yield stress, causes dislocation motion and generates a micro-crack. The micro-crack grows and propagates very slowly because of the low level of stress, but the extended crack eventually causes failure, which is referred to as fatigue failure.

Appendix L: metal matrix composites in engines

Metal matrix composites (MMCs) are used in high-performance engines. Table L.1 summarizes current examples of MMC technologies in automotive engines. These composites are light and strong, but tribological problems must be taken into account when considering their use in engine design.

References

1. Yamagata H. and Koike T., *Keikinzoku*, 49 (1999) 178 (in Japanese).
2. Yamauchi T., SAE Paper 911284.
3. Donomoto T., *et al.*, SAE Paper 830252.
4. Yamaguchi T., *et al.*, SAE Paper 2000-01-0905.
5. Ushio H. and Hayashi N., *Keikinzoku*, 41(1991) 778 (in Japanese).
6. Fujime M., *et al.*, *JSAE Review*, 14(1993), 48 (in Japanese).
7. Gerard D.A., *et al.*, M.C. Fleming Symposium, (2000).
8. Hayashi N., *et al.*, *Nippon Kinzokugakkai Kaihou*, 25(1986) 565 (in Japanese).

Table L.1 MMC technologies in automotive engine parts

Engine parts	Material	Manufacturing process	Manufacturer of the final products
Piston	Extruded PM-Al alloy	Forging	Yamaha[1]
Piston: reinforcement of the piston head and top ring groove	SiC whisker or aluminum borate whisker+High-Si Al cast alloy	Cast-in by squeeze die-casting	Suzuki[2]
Piston: reinforcement of the top ring groove	Alumina fiber +High-Si Al alloy	Cast-in by squeeze die-casting	Toyota[3]
Exhaust valve	Extruded PM-Ti alloy	Forging	Toyota[4]
Cylinder bore	Some types	Cast-in by squeeze die-casting	Several
Connecting rod	Stainless fiber +Al cast alloy	Cast-in by squeeze die-casting	Honda[5]
Crankshaft pulley	Alsilon fiber +Al-12Si-1Cu-1Mg alloy	High-pressure die-casting	Toyota[6]
Engine subframe	Alumina particulate+A6061	Extrusion of cast ingot	GM[7]
Valve spring retainer	Alumina, zirconia and silica particulates +PM-Al alloy	Extrusion	Honda[8]

Index

abnormal microstructures 181, 186–7
AC4B alloy 162–3
AC8A alloy 59, 60–2, 63, 77, 80
AC9B alloy 59, 63, 64, 77, 80, 81, 82, 83
accommodation 105
adhesive bonding 297, 298
AFP1 PM alloy 59, 81, 82, 83
age hardening 36, 261, 291, 302–3
    piston 68–70, 77
ageing 261
age softening 68–70
    and piston temperature during operation 71–2
air-cooling 10, 13–14
air/fuel ratio 229–30, 231–2, 233, 238
air hammers 175
air pollution 228
    see also carbon monoxide; catalysts;
        hydrocarbon; NOx
alloys 261
aluminum
    casting 299–304
    and catalyst honeycomb substrate 235
    coating on outside surface of liner 28–9
    PM aluminum liner 31–2
    stress-strain curve 305–6
aluminum alloys
    Al-Cu alloy 68–70
    Al-Si alloys *see* silicon
    Al-Sn-Si alloy 222
    casting 299–304
    lightweight connecting rods 226
    pistons 57–65, 79
aluminum blocks 15, 16
    casting 26, 40–6
    with enclosed cast iron liners 25–9
ammonia 245, 246
annealing 177–8, 261, 283, 285, 288

low-temperature annealing 262, 307
anodizing 73, 74, 294
articulated piston 83
assembled camshafts 118, 126–8
assembled connecting rods 209, 210, 218–21
    con-rod bolt 218–21
    plain bearing 222–4
    structure and material 218, 220
assembled crankshafts 165, 166, 169
austempering 261, 283, 288
austenite 261
    iron-carbon phase diagram 269–75
    retained 181, 187, 215, 287–8
austenitic nodular cast iron (Niresist) 75, 76, 255,
    256, 259, 278
austenitic steel 134, 136–9, 255, 259

back-pressure, reducing 239
backing metal 222–3
bainite 195, 197, 198, 261, 288
beach mark 76
bearing life 215–17
big end 178, 207, 209–11, 218, 220
bolt, con-rod 218–21
bonded camshaft 118, 126–8
bonded valve 139–43
bonding
    diffusion bonding 128, 235
    by rolling 223
    technologies 128, 297–8
    see also welding
bore interval, shortening 32–3
boring 125–6
brazing 297, 298
bucket tippet *see* valve lifter

cam lobe 113, 133
    improving wear resistance of 116–28